Acoustics, Noise and Buildings

ACOUSTICS

Active Leadership in
Enterprise and Engagement

AND

P. H. PARKIN

Senior Principal Scientific Officer
Building Research Station, Ministry of
Public Building and Works

AND

H. R. HUMPHREYS

Consultant Architect

WITH A FOREWORD BY
HOPE BAGENAL

FABER AND FABER

First published in 1958
by Faber and Faber Limited
3 Queen Square, London, W.C.1
First published in this edition 1963
Reprinted 1966
New and revised edition 1969
Reprinted 1971
Made and printed in Great Britain by
William Clowes & Sons, Limited
London, Beccles and Colchester

© *Peter Hubert Parkin and*
Henry Robert Humphreys
1958, 1969

ISBN 0 571 04672 X (*Faber paper covered edition*)
ISBN 0 571 04671 1 (*hard bound edition*)

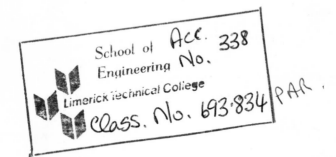

Acknowledgements

We are glad to make these acknowledgements.

The chapter on the design of rooms for music has been strongly influenced by the discussions we have had during the past few years with Mr. Hope Bagenal, and although the views expressed here are not necessarily his, we are most grateful to him. The book *Planning for Good Acoustics* which he wrote with Dr. Alex. Wood (unfortunately now out of print) is still—a quarter of a century later—in our opinion very well worth consulting.

Some of this book, particularly Chapters 9 and 10 which deal with noise measurements and criteria, is based partly on the information given in the *Handbook of Acoustic Noise Control* (Parts I and II and Supplements), prepared by the consulting firm of Bolt, Beranek and Newman on behalf of the U.S. Air Force.

The chapters on the design of rooms for music, on speech reinforcement systems and on sound insulation draw largely from the work of the Building Research Station. The authors also have to thank the Controller, H.M. Stationery Office, and the Director of the Building Research Station for permission to reproduce Plates X and XIII from unpublished photographs taken by the Building Research Station, and to use information relating to experiments in Salisbury Cathedral, given on pages 164 to 166, which is taken from unpublished records at the Building Research Station. Crown copyright in these unpublished photographs and information is vested in the Controller, H.M. Stationery Office.

The chapter on the design of studios is largely based on experience gained with the British Broadcasting Corporation, to whom we are indebted for Plates V, VI, VII and IX.

We also wish to thank the following, for kindly supplying the undermentioned illustrations: *Architectural Design*, Fig. 28; The Greater London Council for Plates I and III; The U.S. Information Service for Plate IV; Anderson Construction Co., Plate VIII; Telefunken Ltd. for Plate XI; Treetex Acoustics Ltd., Plate XII; Brüel & Kjaer for Plate XIV, and the *Architects'*

ACKNOWLEDGEMENTS

Journal for Fig. 24. Figs. 23, 26, and 31 are drawn from information published in *Acoustical Designing in Architecture* by Knudsen and Harris, published by John Wiley & Sons, New York; Chapman and Hall (London).

Finally, we wish to thank Margaret Humphreys for considerable help in the preparation of the drawings which illustrate this book.

PREFACE TO SECOND EDITION

Advantage has been taken of the first reprinting to make some minor corrections, but major alterations have been limited to two items. The first is a downward revision of permissible noise levels to avoid permanent deafness (pp. 287–288). This revision is based on very recent work, and in view of the importance to health we felt the new figures should be given as quickly as possible. The second item concerns sound level meters (pp. 262–263) which in the first printing were dismissed as seldom giving the true loudness of a noise. This is still so if they are used as originally intended, but used in a different way they can provide a simple and reliable method of measuring the loudness of many types of noises.

PREFACE TO THIRD EDITION

The principal change incorporated in this edition is the conversion of all units to the metric system to fall in line with the adoption of these units in the construction industry. Those not yet familiar enough with metric dimensions can relate them to Imperial units by reference to the conversion table on p. 324.

Slight revisions have been made in the text where current thought or disciplines have modified the original writing.

Foreword

Building acoustics, this 'difficult but fascinating subject' as the authors of this book call it, demands a combination of studies. It demands a knowledge of physics together with a thorough familiarity with building techniques. Planning and town planning enter it. It needs—for concert hall and music studio—a study of musical tone and musical taste. For speech requirements it must envisage modern speech reinforcement and the loudspeaker. This means that a number of experts must take part together with architect and physicist. It is a border-line subject lying between the great principalities of modern knowledge, and suffers accordingly. Width of experience, visits to halls, field tests, are necessary; and in fact there are very few scientists who make it a whole-time study; and as a result of these contingencies text-books are few. At the same time large new auditorium buildings of all kinds—legislative chambers, churches, assembly halls, theatres and (specially difficult) radio studios—are now being proposed and built in all the countries of the British Commonwealth. And such buildings involve not only room acoustics. We must also defend them against the immensely increased risk of disabling noise—noise from aircraft, from street traffic, from new industrial noises, from mechanical equipment indoors. This need for a civilisation to defend itself against chains of distraction hindering mental activity is brought home to designers at every turn. And it will force itself more and more upon public attention as air traffic grows, and main runways increase in length, and aircraft of all kinds get faster and come in over residential areas at a lower angle of approach. Hence in this book an equal importance is given to insulation and to room acoustics. In both aspects the authors have embodied the results of the many valuable investigations made, since the war, by the Building Research Station. But in addition to their knowledge of tests and experiments they have considerable practical experience so necessary for discriminating the things that matter in a modern building under pressure of a thousand conflicting requirements. They have built and they have listened. And so I hope that their broad survey of the whole field will refresh the student and help to define English practice, and that at the same time the impartial reviews they give of speech reinforcement systems, of television studio require-ments, of concert platform problems, will prove specially valuable to designers in that acid test of modern theory, namely usefulness for practical specifying. The authors hope they have given us, in their own words, 'the best practical advice in the light of modern knowledge'.   That is the kind of book that will help us in the kind of problems which now stand before us.

HOPE BAGENAL

Introduction

Probably the first reference to building acoustics in recorded history is that by Vitruvius, to the sounding vases or echeia which he says were used in ancient Greek open-air theatres, but the influence of acoustics on the design of buildings can be observed all through the ages. Extensive quantitative study of the subject had to wait until the nineteenth century when, among others, W. C. Sabine did his pioneer work on the practice of acoustic design, Lord Rayleigh wrote his classical exposition, *The Theory of Sound*, and Helmholtz developed the theory of music. To-day wide new fields of study have been opened up by the work of physicists, musicians, engineers and broadcasting and sound-film technicians. Although the closer investigation of sound-wave theory has led physicists into mathematics too complex for practical workers to use, these studies have been accompanied by great strides in acoustical engineering.

The development of broadcasting and recording has given us new ways of examining acoustical phenomena, presented us with new problems to solve, and awakened a widespread popular interest in listening to music. For example, one problem which these techniques has solved is that of the production of intelligible speech from the pulpit or lectern to the whole of the congregation in churches with very long reverberation. This difficulty has probably become more obvious in recent years but the English Common Prayer Book (sixteenth century) indicates that it may be of long standing. It states: 'Then shall be read distinctly with an audible voice the First Lesson. . . . He that readeth so standing and turning himself as he may best be heard of all such as are present.'

But another problem of great practical importance faces us with increasing insistency to-day. That is the problem of noise, which has also been with us for a long time. In 1682 the following advertisement appeared: 'At one Mr. Packer's in Crooked Lane, next the Dolphin are very good lodgings to be let, where there is freedom from noise and a pretty garden.' Or we can go

9

even farther back. R. E. Wycherley writes: 'The Athenian Pnyx [assembly place] was a short distance to the south-west of the classical agora (or central market and public space) of Athens. At first the natural hillside was used, with perhaps a natural platform for speakers at its foot. Towards 400 B.C. the whole building was turned round. A curved retaining wall was built at the bottom and within this a great embankment was raised in order to give the auditorium a slight upward slope (now in the reverse direction to the hillside). The motive for this curious reversal is obscure ... the Greek architects seldom went so strongly against nature in such buildings.' But were they going against nature? We suggest that their reason was the difficulty of providing the necessary degree of quietness for intelligibility in the meeting place when the busy market of an expanding town gave rise to so much noise. So man has always taken steps to control his environment and must continue to do so or suffer the consequences.

To-day the consequences may be severe. Anything which works against our finer perception and awareness of the physical world around us we count as an anti-civilising influence. It cannot be said that progress has been made if in the process some of our senses have had to be blunted to make a new environment acceptable. Everyone has, at some time or another, been annoyed by noise, but as at present annoyance is not a strictly measurable function, we must be content with examining what actual physiological harm noise can cause. The study of the overall effects of noise by psychologists continues, but their work, and that of sociologists and lawyers on the same subject, is dogged by the diffuseness of the 'average' man and the obtrusiveness of the hypersensitive one. Leaving aside these finer points, the fact remains that to-day some people are being permanently deafened by noise, while many others have their work, pleasure or sleep seriously disturbed.

To define noise is difficult and probably the internationally accepted definition 'sound undesired by the recipient' is the closest we can hope to get. Most people would have no hesitation in placing the sound of an experimental jet aircraft returning to land in the category of noise, but the designer of the aircraft and the test pilot's wife are unlikely to hold this opinion. Jet aircraft are an example of the technological developments which are creating both the supply of problems and the demand for their study. As one of James Thurber's dogs has cleverly noted, 'Men built strong and solid houses for rest and quiet and then filled them with lights and bells and machinery.'

The influence which both of these aspects of our subject has on building, and vice versa, is too obvious to need long discussion. We will confine ourselves to suggesting that the technical problems presented demand technical solutions.

These problems are the concern of a very wide group of people. The late Alexander Wood, in his inaugural address to the first summer symposium of the Acoustics Group of the Physical Society in 1947 said: 'The tower of human knowledge which we are in process of building is endangered by the curse of Babel and many who are engaged in its construction have ceased to be intelligible to any of their fellow workers except those actually working on the same part of the building. . . . Although it is necessary for us all to talk our specialist language to our specialist colleagues, it would do us all good if we occasionally placed ourselves under the discipline of explaining to a wider circle what we are trying to do.' And this is exactly what we have tried to do in this book. To this end we have eschewed all complex mathematics and have avoided, as far as possible, digressions on aesthetics. We have not adhered consistently to standard scientific terms but have employed those which are in common usage.

This book is not, then, a textbook on physical acoustics in the normally accepted sense, nor is it addressed only to architects or others solely concerned with buildings. The information is designed for specialist readers of a number of kinds, and may even be of some use to laymen. Chapters 1 to 5 and 7 and 8 contain the information which concerns architects and students of architecture and building; Chapters 6, 8 and 9 are primarily for the acoustic engineer, but he may find some of the earlier chapters worth reading. The broadcasting engineer will be concerned with Chapters 5 and 6, but there is also information about the sound insulation of studios in Chapter 8. Lastly, administrators, directors and town planners will find something to their interest in Chapters 1, 7, 8 and 10.

In conclusion we have tried to give what we believe to be the best practical advice in the light of present knowledge of this difficult but fascinating subject. New developments are constantly occurring and the periodical literature which can be studied includes, on physical acoustics, *The Journal of the*

Acoustical Society of America, Acustica, and *Journal of Sound & Vibration.*

For further reading, the following publications are suggested. The technique of acoustic measurements—only briefly touched on here—is fully described in *Acoustic Measurements* by Beranek, published by John Wiley & Sons, Inc., New York (Chapman & Hall Ltd., London). *Noise Measurement Techniques,* prepared by the National Physical Laboratory and published by H.M.S.O. will also be found most useful. *Handbook of Noise Control,* edited by C. M. Harris and published by McGraw-Hill (New York) deals in considerable detail with the engineering control of noise. The British Standard Code of Practice CP.3: Chapter III (1960) *Sound Insulation and Noise Reduction* gives general guidance on the design of buildings. Research Paper 33 in the National Building Studies, *Field Measurements of Sound Insulation Between Dwellings,* prepared by the Building Research Station and published by H.M.S.O. deals in detail with the sound insulation of party walls and floors. Details of the sound absorption coefficients of many materials are given in *Sound Absorbing Materials,* prepared by the National Physical Laboratory and published by H.M.S.O.

"The second point is Usefulnesse, which will consist in a sufficient Number of Roomes, of all sorts, and in their apt Coherence, without distraction, without confusion."

The Elements of Architecture, SIR HENRY WOTTON (1624)

Contents

CONTENTS

Illustrations

I

Nature of Sound

Sound is the sensation produced through the ear resulting from fluctuations in the pressure of the air. These fluctuations can be set up in a number of ways, but usually by some vibrating object, and are in the form of a longitudinal wave

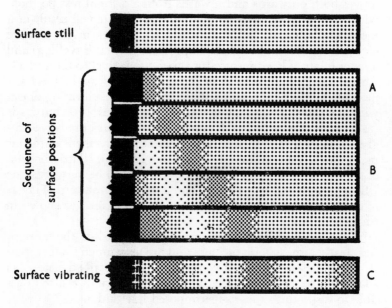

FIG. 1 (A, B, C). Illustration of Generation and Propagation of Plane Sound Waves

motion. Consider the air close to the surface of some object which is vibrating. As the surface moves outwards the air molecules next to the surface are pushed closer together, i.e. the air is compressed. The air cannot move back into its original position

23

for the moment because the space is occupied by the advanced surface of the vibrating object and therefore a movement of air occurs away from the surface. This movement in turn causes the compression of another layer of air. This is followed by a further release of pressure, again by movement of air outwards from the vibrating surface. The result of the outward displacement of the vibrating surface is thus to produce a layer of air compression progressing outwards from and parallel to the surface.

This does not mean that air travels as a whole outwards from the vibrating object. It is only the wave of compression which travels continuously as a result of small limited movements of air molecules in this direction (see Fig. 1A). The wave of pressure will move outwards at a steady rate and after it has gone a certain distance, i.e. after a certain time, the surface of the vibrating object will have moved in again through its rest position to one further back. (It is assumed that the surface moves both outwards and inwards from a normal rest position, which is the usual behaviour.) This movement will result in a movement of some air molecules back again (in the reverse direction to the one in which the pressure wave is travelling) and a layer of air of low pressure (or rarefaction) next to the surface. Again, another movement of the molecules of air in the layer next further from the surface back towards the vibrating object will take place. This small movement will also progress outwards (as a chain of events) in the same way as the pressure wave progressed outwards, although in this case the movement of the air molecules is in the reverse direction (see Fig. 1B). If the vibration of the surface continues its next movement will be outwards, the whole process will be repeated, and a travelling pattern of layers of compression and rarefaction will be established as shown in Fig. 1c.

These pressure fluctuations are superimposed on the more or less steady atmospheric pressure and are very much smaller than it. Nevertheless the ear is constructed (as described below) so that it is not sensitive to the steady atmospheric pressure but is able to distinguish these superimposed fluctuations.

The speed of travel of this pattern (in air at atmospheric pressure and temperature of 14°C) is about 340 m per second independent of the rate at which vibrations are being produced. The distance between the layers of compression, i.e. *the wave-length*, will obviously depend on how fast the vibrations which cause them are taking place. If, for example, one cycle (one complete in and out motion) occurs in one hundredth of a second, then the wavelength will be $\dfrac{340}{100} = 3 \cdot 4$ m. The rate at which the vibrations occur is called the *frequency*, usually stated in cycles per second, and the constant relationship which exists between these quantities is:

$$\text{wavelength} \times \text{frequency} = \text{speed of sound}$$

The diagrammatic representation of Fig. 1 shows the sound wave travelling in one plane and this is called a 'plane wave'. In practice we are more concerned with what are called 'spherical waves' in which the sound waves travel out from the vibrating object—the source—in every direction. Thus a closer approximation to many practical noise sources would be to replace the plane surface in Fig. 1 by a spherical body which is periodically expanding and contracting. The waves of compression and rarefaction will travel out from the source in a similar manner but they are now spherical and an important practical consequence is that their intensity (i.e. the energy per unit area of the sound wave) will diminish with increasing distance from the source. The reason is that the energy imparted to the air by the source is spread over a greater area as the distance increases, and as the area of a sphere is proportional to the square of the radius it follows that the intensity of the sound waves will decrease inversely with the square of the distance from the source. This is known as the inverse square law and will be referred to later.

When a sound wave meets an obstacle its behaviour will depend on the nature of the obstacle and on its size relative to the wavelength of the sound. This, and the behaviour of sound waves in rooms, is discussed in the following chapters.

Sound sources in practice mostly consist of some vibrating element. For example, in the human voice the air from the lungs is forced past the vocal cords which vibrate and allow an intermittent flow of air to reach the vocal cavities. After modification by these cavities the sound waves are radiated from the mouth. The violin string vibrates when the bow is drawn across it, but the string itself radiates very little sound because it is so small. However, the way in which it is coupled to the body of the violin causes the body to vibrate and this radiates the sound.

FREQUENCY RANGE

The frequencies we are interested in in this book are those to which the human ear responds and the range is from about

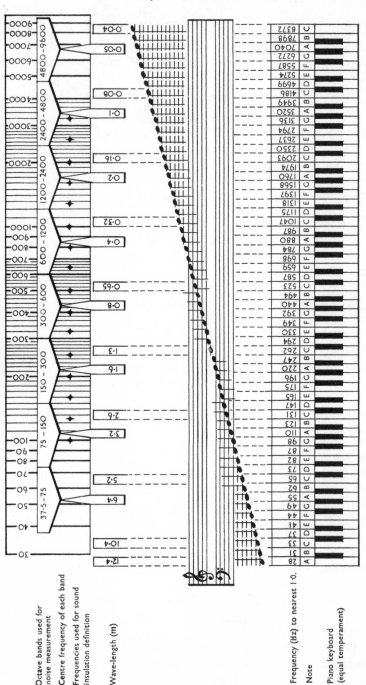

Fig. 2. Frequency Ranges

30 cycles per second or Hertz (abbreviated Hz) up to about 20,000 Hz, although frequencies higher than 10,000 Hz are seldom important. The corresponding wavelengths are from about 11·3 m down to about 12 mm. Fig. 2 illustrates this range of frequencies as related to a piano keyboard. The frequency is plotted on a logarithmic scale because the pitch of a sound as heard by the ear varies in this way. For example, every time the frequency of a tone is doubled the pitch goes up one octave. Thus the pitch difference between 100 and 200 Hz is one octave and so is the difference between 1000 and 2000 Hz On a linear frequency scale the space between 1000 and 2000 Hz would be ten times that between 100 and 200 Hz, whereas on the logarithmic scale the two spaces are equal, to correspond to the pitch change. (Actually, the pitch of a note also depends to a small extent on its loudness, but this need not concern us here.)

The frequencies shown next to the keyboard on Fig. 2 refer to the fundamental (or first harmonic) for each note. However, every musical sound contains in addition to its fundamental a series of over-tones (or harmonics) which are multiples of the first harmonic. Thus the second harmonic is twice the frequency of the first harmonic, the third harmonic is three times the frequency, and so on. It is the number and relative strength of the harmonics which distinguishes one musical instrument from another (at least when the instrument is playing a steady note). One of the 'purest' wave-forms, i.e. with little harmonics, is produced by the flute, while the 'richness' of piano tone is due to the large number and comparatively high intensity of its harmonics.

Another difference between musical instruments is their transient response, i.e. the way they behave when first struck or blown. For example, a piano never produces a steady note because the string is either just being struck or is decaying in intensity after being struck. On the other hand, an organ pipe has an initial transient period when it is first sounded but thereafter will continue to emit a steady note.

The Ear

A diagram of the ear is given in Fig. 3, where it is seen that the ear-drum is in contact with the air through the auditory canal (called the outer ear) which is about 25 mm long. The atmosphere is pressing on the ear-drum with a force very much greater than the force any sound wave will produce, and if there were no compensatory mechanism the ear-drum would be pushed hard in. But from the inside of the ear-drum (called the middle ear) the Eustachian tube connects with the mouth and allows the atmospheric pressure to operate on the inside of the ear-drum also. The result is that the ear-drum is not affected by the atmospheric pressure nor by the slow changes which occur in it as the weather changes. The Eustachian tube operates comparatively slowly (and only during the action of swallowing) so that if there is a sudden quick change in the atmospheric pressure, as happens in a rapid descent in an aircraft, then it

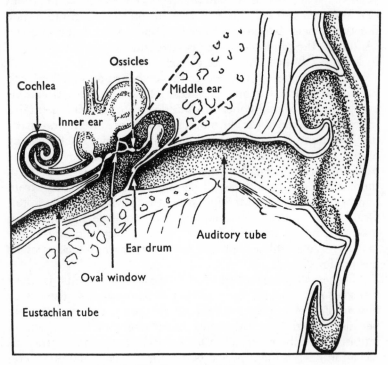

Fig. 3. The Ear and its Principal Parts

may not work quickly enough and the ear will hurt. If the Eustachian tube operated instantaneously, i.e. if the inside of the ear-drum were connected directly to the atmosphere, then not only would the atmospheric pressure be equalised but the sound pressures would also, and so the ear would not work.

The alternating sound pressures push the ear-drum in and out, and this movement, which like all movements and forces involving sound is very small, is transferred by three small bones (the ossicles) joined together (and called the hammer, anvil and stirrup) to a second membrane, the oval window. This second membrane separates the middle ear, which is filled with air via the Eustachian tube, from the inner ear. The mechanical advantage given by these small bones and the fact that the oval window is much smaller than the ear-drum combine to increase the sensitivity of the ear.

The oval window is at one end of the cochlea, a hollow member made of bone and filled with liquid. It is spiral shaped with an unwound total length of about 38 mm. It is divided down the middle by the cochlea partition and part of this partition consists of the basilar membrane on which terminate about 25,000 endings of the main auditory nerve.

Thus an alternating pressure in the air acts first on the ear-drum and this in turn acts on the oval window through the action of the ossicles; finally the pressures on the oval window are transmitted to the liquid in the cochlea. The resulting alternating pressures in this liquid are detected by the nerve endings and transmitted to the brain as the sensation of sound.

Decibels, Phons and Sones

The range of sound pressures which the ear responds to is very large although the pressures themselves are small. For example, at 1000 Hz the threshold of audibility (i.e. the quietest sound that an average person can just hear) occurs at a sound pressure of about 0.0003 dyn/cm^2, while the pressure at which the ear begins to hurt is about a million times as great—of the order of 300 dyn/cm^2. (For comparison, the atmospheric pressure is about 1,000,000 dyn/cm^2.) If sound pressures were given in dyn/cm^2 we would have a very large and awkward range of numbers to deal with. Further, subjectively perceptible increases in loudness are not equal for uniform increases in sound pressure. It is as though the perceived difference in two lengths, one of 2 m and the other of 3 m, was quite at variance with the perceived difference between another two lengths, one 20 m and the other 21 m. In order to obtain roughly equal perceptible differences in sound level the ratio between the two sound pressure quantities must be constant throughout the scale. Thus, to get the same subjective difference between two pairs of sounds, one at 2 and the other at 3 units, the second pair would need to be 20 and 30 units, i.e. so that the second of the two pairs of values has the same ratio to one another (2:3) as the first pair.

This is the basis of the decibel scale which is used to reduce the range of numbers and because it roughly fits the human perception of the loudness of sound.

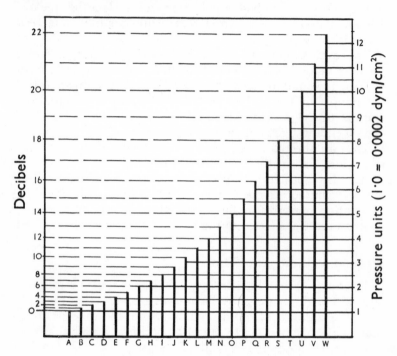

FIG. 4. Relation between Sound Pressures and Decibels

The illustration (Fig. 4) shows the relation between sound pressures and decibels. This scale is logarithmic (as compared with linear such as a foot rule). For example, a one decibel increase corresponds to an increase in sound pressure by a factor of about 1·1 (more precisely 1·122 . . .).

The decibel is not an absolute measure but always a relative one, i.e. it always gives the ratio of two pressures. For our purposes it will be sufficient to define the decibel as 20 times the logarithm (to the base 10) of the ratio of the two pressures, i.e.

$$\text{decibel ratio} = 20 \log \frac{p_1}{p_2}$$

where p_1 and p_2 are the two pressures being compared. For example, if one pressure is twice the other the decibel ratio is 20 log (2/1) = +6 dB. Some commonly occurring ratios are: a factor of 2 = 6 decibels (abbr. dB), and it follows that a factor of 4 = 2 × 2 = 6 + 6 dB = 12 dB, a factor of 8 = 18 dB, and so on; a factor of 10 = 20 dB and thus a factor of 100 = 10 × 10 = 20 + 20 dB = 40 dB, and so on up to, for example, the pressure range we have mentioned of 1,000,000, which equals 120 dB. The larger pressure of the two being compared is so many dB greater than the other, e.g. a pressure of 2 dyn/cm² is 6 dB greater than a pressure of 1 dyn/cm². It can also be put the other way round, i.e. a pressure of 1 dyn/cm² = a pressure of 2 dyn/cm² minus 6 dB.

Sometimes it is only the relative pressures that are required. For example, if sound insulation is involved the pressure on one side of a wall might be, say, 20 dyn/cm² (at a particular frequency) and on the other side this might be reduced by the wall to 0·2 dyn/cm². The ratio of the two pressures is 100, or 40 dB. If the first pressure were to be doubled the pressure on the far side of the wall would also go up double and the ratio would still be the same. Thus the pressure reduction due to the wall, or in other words its sound insulation, is always 40 dB no matter what the absolute values of the two pressures are.

However, it is often the absolute value of a pressure that is required and then it is necessary to have a reference pressure if dB's are to be used. There is an agreed reference pressure and it is 0·0002 dyn/cm² (i.e. a little below the threshold of hearing at 1000 c/s). Thus if the noise of, say, a pneumatic drill is measured and found to be 20 dyn/cm² then this is a factor of 100,000 greater than the reference pressure, i.e. it is 100 dB greater. The noise is then described as being at a level of 100 dB and when noise levels are given as being so many dB this always means so many dB greater than the reference pressure.

The ear is less sensitive at the lower and higher frequencies (i.e. it takes a greater alternating pressure to produce the same sensation of loudness) than it is at the medium frequencies. However, this difference in sensitivity also varies with how loud the sounds are: only at high sound pressures is the ear roughly equally sensitive to all frequencies.

As we have said, at 1000 Hz the ear can just detect a sound pressure of about 0·0003 dyn/cm² (or 4 dB above the reference pressure). At 50 Hz a greater pressure is required before the ear can detect it—about 0·02 dyn/cm² (41 dB). At 3000 to 5000 Hz the ear is a little more sensitive than it is at 1000 Hz and will detect a pressure of about 0·00015 dyn/cm² (−3 dB). At the higher frequencies its sensitivity drops; for example, at 10,000 Hz it can detect a pressure of about 0·0015 dyn/cm² (17 dB).

This threshold of audibility is shown in Fig. 5 as a function of frequency. Also shown is the upper limit of hearing, i.e. the levels where the noise is so loud that it begins to hurt—'the threshold of pain'. Between these two limits occur all the pressures we are now concerned with.

We must now deal with how loud a noise of a given frequency and of a given pressure will sound. We have seen that the threshold of audibility is different at different frequencies and similarly the loudness of a sound depends on both its pressure

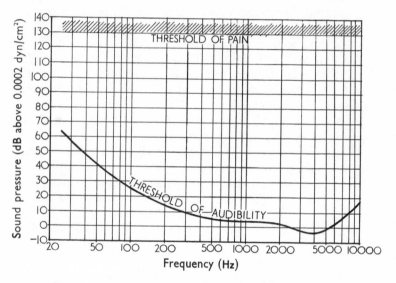

Fig. 5. Threshold of Audibility and Threshold of Pain

and its frequency. (For the moment we are considering only pure tones, i.e. of one frequency only, which are unlike most sounds and noises occurring naturally or by simple human agency which are made up of many frequencies.) To form a basis of comparison pure tone of 1000 Hz is used as a reference and a pressure of 0·0002 dyn/cm² (0 dB) is the zero on a scale of units called *phons*. At 1000 Hz the phon level is the same as the decibel level, e.g. a sound pressure of 200 dyn/cm², i.e. 120 dB, has a loudness of 120 phons.

(It should be realised that this definition of the phon is quite arbitrary; it has no particular physical or physiological basis but is merely a convenient method, as will be shown, of providing a basis for standards of loudness.)

At other frequencies than 1000 Hz the ear's sensitivity is different and its judgement of loudness at these frequencies is compared with its findings at 1000 Hz. This comparison must of course be done subjectively and it is usual to use a large number of people who are known to have no defect in their hearing, and to take the average of all their results as being typical of the 'normal' person. This comparison has been done in great detail by many laboratories at various times and

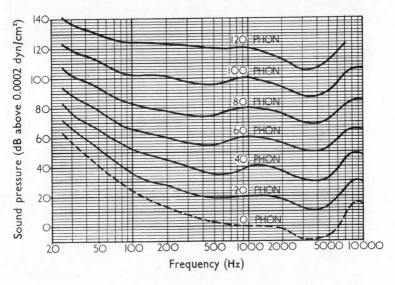

Fig. 6. Equal Loudness Contours

although minor discrepancies, some of them unexplained, have appeared the general agreement between results is good.

We have defined the loudness of a tone at 1000 Hz and at a pressure of, say, 60 dB as being 60 phons, so it follows that if some other frequency tone is adjusted to sound as loud as the 60 phon 1000 Hz tone its loudness must also be 60 phons. Thus a 100 Hz tone at a pressure of 66 dB is found to have a loudness of 60 phons, and a 3000 Hz tone at a pressure of 51 dB also has a loudness of 60 phons.

This sort of comparison has been made at all frequencies and over the whole pressure range the ear is sensitive to, and the results are usually given as in Fig. 6 (after Robinson and Dadson). The vertical ordinate shows the pressure levels (in dB greater than 0·0002 dyn/cm²); the horizontal ordinate shows the frequency (in Hz): plotted against these two variables are contours of equal loudness. The bottom contour is 0 phons and is a little below the threshold of audibility. The other contours go up in 20 phon steps to the top one of 120 phons.

From this graph any single tone can be transposed from dB's into phons, or vice versa. For example, if we have a tone at 100 Hz at a pressure of 44 dB then we can fix this point with reference to the two ordinates and it is seen to come halfway between the 20 and the 40 phon contours. This tone, then, will have a loudness of 30 phons. Similarly, we could say that a tone of 50 Hz at a loudness of 80 phons must be at a pressure level of 94 dB.

It has been mentioned earlier that the difference in the sensitivity of the ear to different frequencies is greatest at low pressure levels, and this is seen in Fig. 6. At 50 Hz a pressure of 47 dB is required to produce a loudness of 10 phons, as compared with 10 dB at 1000 Hz, i.e. a difference of 37 dB for the same loudness; at 50 Hz a pressure of 112 dB is required to produce a loudness of 100 phons, as compared with 100 dB at 1000 Hz, a difference of 12 dB.

Arising out of this it should be noted that a change of pressure level at the low or at the high frequencies produces a greater change in phons, i.e. in loudness, than the same change in pressure level would do at the mid-frequencies. Thus reducing a pressure level of 80 dB at 100 Hz to 70 dB reduces the loudness from about 77 phons to 63 phons, while the same change—80 to 70 dB—at 1000 Hz reduces the loudness by 10 phons.

The smallest change in loudness of sounds of moderate loudness that the ear can detect is of the order of 0·5 dB. This is so only when the loudness is changed instantaneously or with only a short time interval between changes and it certainly does not mean that such small changes could be detected in practical cases, e.g. when an intruding noise has been reduced by improving the sound insulation. What is the smallest detectable difference under these conditions is not known but it appears that at least under some conditions a 5 dB change in noise levels represents a definite and permanent subjective change. One example is the 5 to 6 dB difference between the two grades for sound insulation (see p. 297) and this equals a subjective change of from minor to considerable annoyance. Another example is the criterion for estimating community reaction (see p. 303), where the subjective steps correspond to 5 dB differences.

Another unit is sometimes used for loudness and this is the sone. The scale is such that the number of sones is proportional to the loudness, i.e. a noise of loudness $2x$ sones will be twice as loud as a noise of x sones. The relation between phons and sones

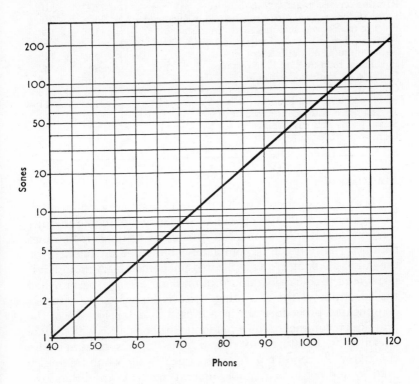

FIG. 7. Relationship between Phons and Sones

Note that the relationship plotted here is slightly different, at the higher levels, from the relationship given in Chapter 9, p. 264, and from which the loudnesses given in Table I were calculated. The reason is that these two relationships were determined by different laboratories.

(over the range 40 to 120 phons) is shown on Fig. 7. It is seen that halving or doubling the loudness of a noise equals a change of about 10 phons.

The loudnesses of some common noises is given in Table I both in phons and in sones.

TABLE I

LOUDNESSES OF COMMON NOISE

Noise	Phons	Sones
Large jet air-liner when 38 m overhead	140	1020
Riveting on steel plate at 2 m distance	130	510
Large piston-engined air-liner when 38 m overhead	130	510
Electric trains over steel bridge at 6 m distance	110	130
Weaving shed	100	64
Steam trains at 30 m distance	95	45
Pneumatic road drill at 38 m distance	90	32
Heavy road traffic at kerb-side	85	22
Canteen	80	16
Male speech at 1 m distance	75	11
Typing office with acoustic ceiling	75	11
Steam trains at 2 km distance	65	6
Light road traffic at kerb-side	55	3

MASKING

Masking is concerned with the effect of one noise on another. For example, while speech will be perfectly intelligible in quiet surroundings it will become less intelligible as the surroundings get noisier until it will be completely obliterated or 'masked' by the noise. Masking is a complicated phenomenon. Although it is possible to predict the amount of masking of, say, speech when the spectrum of the intruding noise is known, it is a rather complicated procedure and is not used in detail for any of the practical recommendations made in this book. We will confine ourselves to pointing out that a pure tone is masked more effectively by another pure tone of nearly the same frequency than by another tone of quite different frequency, and that low-frequency noises mask high-frequency noises more than high-frequency noises mask low-frequency noises.

HEARING

The structure and response of the ear have already been briefly described. We give now a short outline of audiometry and of the more common types of hearing defects. It is intended to serve as a rough guide only (for example to those who may be concerned with the hearing loss of factory workers). A reliable assessment of a person's hearing can only be got by a proper clinical examination.

AUDIOMETRY

Audiometry is the term used to describe the measurement of hearing loss, and the instrument used is called an audiometer. This instrument produces pure tones at various frequencies and at pressure levels which can be adjusted over a wide range. The subject wears earphones and, as the level of one of the pure tones is raised, is asked when he can just detect it. If he has no hearing defects then the pressures indicated by the audiometer will be as the bottom curve of Fig. 5. A person who has some hearing defect will not be able to hear these tones at these pressures and the tones must be made louder before the deaf person can hear them. The amount by which the levels must be raised above the normal threshold is defined as the 'hearing loss'. For example, a person suffering severely from otosclerosis (see below for discussion of types of deafness) may only just be able to hear at 200 Hz a level of, say, 76 dB compared with the 'normal' 14 dB, i.e. he has a hearing loss of 62 dB at this frequency; at 1000 Hz the level may have to be 49 dB before he can just hear it, compared with the normal 4 dB, i.e. a hearing loss of 45 dB; and so on at all other frequencies. A graph showing a person's hearing loss as a function of frequency is called an audiogram; the audiogram for the above example of severe otosclerosis is shown in Fig. 8.

It is obvious that when an audiometer is being used the noise level in the room is most important. Although the subject will be wearing earphones for the test, noise may still be loud enough to get through the earphones and 'mask' (i.e. make it more difficult to hear) the test tones. Thus the test tones may have to be louder before the patient can detect them than they would have to be in quieter conditions; the hearing loss measured under noisy conditions may be higher than the true hearing loss. When general practitioners are using an audiometer they probably make their own adjustment as best they can from their own experience for the local noise conditions. Hospitals, however, will usually have a special room in which the noise level is low enough not to interfere with the measurements. The maximum permissible noise levels for such audiometry rooms are discussed in Chapter 10 and how to construct the rooms in Chapter 8.

Another type of audiometric test, not so widely used, is to present recorded speech to the patient via earphones at various loudnesses and to measure the articulation. This is a more complicated procedure than pure-tone audiometry but has the advantage that it may show up the phenomenon of 'loudness recruitment' (see below).

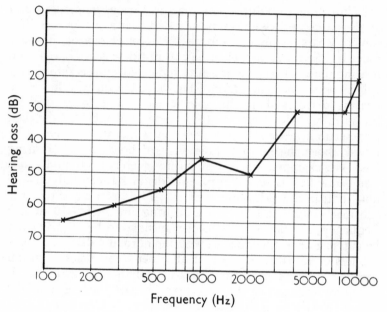

FIG. 8. Audiogram showing severe Otosclerosis

PRESBYCOUSIS

Presbycousis is the medical term for the loss of sensitivity of the ear which occurs naturally with increasing age. There is a complication when we come to decide what is the 'normal' loss due to age and it is that in testing any sample of the population there will be some whose hearing loss will be due not only to age, but will be due also to, say, exposure to loud noises at some period in their life which they may not mention to the tester, or which they may not remember. The older the age group being tested the higher proportion of such people there are likely to be and thus the results may show, on the average, a greater hearing loss than that due to age alone. This may be the explanation of discrepancies between some of the results obtained in the various investigations that have been made. However, it is not a very important effect.

The main characteristic of presbycousis is a loss of sensitivity at the higher frequencies. Fig. 9 gives an indication of the loss to be expected with age for frequencies above 1000 Hz. It is seen for example that at 5000 Hz the loss to be expected at the age of 60 is 17 dB. There is some doubt about the losses at frequencies lower than 1000 Hz. Some recent but not extensive work has shown no losses with age at frequencies up to 1000 Hz, while previous large-scale investigations had shown some losses. However, at the higher frequencies there are undoubtedly considerable losses. It is important when assessing hearing losses due to exposure to loud noises to allow for presbycousis.

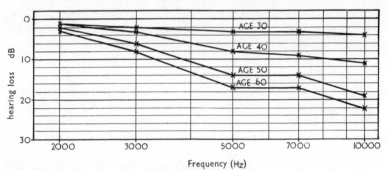

Fig. 9. Presbycousis Loss

Tinnitus

Most people at one time or another have experienced a 'ringing' or some other kind of noise in their ears; this is tinnitus. It appears in many forms, the most common type being a fairly high frequency, not quite pure tone which comes on suddenly, lasts for a few seconds and then dies away. Low-frequency tones are less common but do occur. There are many causes for tinnitus which need not concern us. It is mentioned here because there are occasions when sufferers from continuing tinnitus blame the noise on some external source, such as a factory.

Deafness

The term deafness is used here for the hearing loss due to accidents, disease or exposure to loud noises, and excludes presbycousis.

There are three main types of deafness: conductive, nerve and cortical. Cortical deafness chiefly affects old people, and is due to a defect in the brain centres; it will not be discussed further here. Conductive deafness and nerve deafness may occur singly but may also occur together, providing what is called a 'mixed' type of hearing loss.

Conductive deafness is due to defects in those parts of the ear —namely the external canal, the ear-drum and the ossicles— which 'conduct' the sound waves in the air to the inner ear. Examples of conductive deafness are a thickening of the ear-drum, a stiffening of the joints of the ossicles, or simply a blocking of the external canal by, usually, wax. This type of deafness either affects all frequencies more or less equally, or, more typically, is severer at the low and middle frequencies. It is limited in amount to about 50 to 55 dB (see above for method of measurement of hearing loss) because, even if the normal channel is completely inoperative, some sound will still reach the inner ear by conduction through the bones of the head.

A common form of deafness is otosclerosis, in which a disease of the bone makes the stapes immobile. The deafness increases over a period of years as the disease progresses. It can be alleviated by an operation known as 'fenestration', in which a new window or 'fenestra' is introduced into the lateral semi-circular canal.

A perforation of the ear-drum may be caused either by disease or by accident, e.g. an explosion. This results in some loss of hearing, but often the ear-drum will heal and hearing will return to normal. In severe cases it has recently become possible to replace the damaged ear-drum by an artificial one.

A conductive hearing loss not only raises the threshold of hearing but also affects hearing all the way up the loudness scale. Thus, for example, a person with a conductive hearing loss of 20 dB at 1000 Hz will not only require a sound pressure level of 24 dB at this frequency before he can detect it but will also hear a 60 phon 1000 Hz tone as 40 phons, 80 phons as 60 phons, and so on. One interesting consequence of this is that such a person will have difficulty in understanding speech which is spoken quietly in quiet surroundings, simply because the quieter parts of the speech will be below the listener's threshold of audibility. But in noisy surroundings the speaker will automatically raise his voice to make himself understood: the loudness of both the speech and the noise will be reduced for the listener, but the relative levels, i.e. the amount by which the speech is above the noise, will be unchanged. Since the speech level is now fairly high the intelligibility is not limited by the threshold of hearing but is limited by the level of the noise. Thus a person with a not-too-severe conductive hearing loss may understand speech better in a noisy environment than he does in quiet surroundings.

Nerve deafness (sometimes called perceptive hearing loss) is due to loss of sensitivity in the sensory cells in the inner ear or some defect in the auditory nerve. There is no medical remedy for it. This type of hearing loss is usually different for different frequencies and nearly always is greater at the higher frequencies, i.e. above 1000 Hz, than at the lower frequencies. It thus resembles presbycousis. The hearing losses caused to people working for long periods under very noisy conditions (see below) are of this type.

Some types of nerve deafness show a phenomenon known as 'recruitment'. It is that, while at low loudness levels the person is deaf, at higher levels the hearing is normal. This can be distressing because while such an individual may have difficulty in hearing a person speaking in a normal voice, when the speaker raises his voice he complains that the speaker is shouting.

DEAFNESS CAUSED BY NOISE

If a noise is loud enough it will immediately damage the ear-drum, perhaps so badly that it will not heal, or destroy the delicate ossicles. It is not known with any accuracy how loud a noise must be to cause such serious damage because naturally experiments are not made on human beings, but from experiments on animals and from such accidents to humans that have been investigated it appears to be of the order of 150 dB.

Of more general importance is the hearing loss caused by exposure over periods of time to the noise levels that often occur in industry. Exposure to these noises causes nerve deafness, i.e. damage to the sensory cells of the inner ear, or possibly to the auditory nerve. No surgical remedy is possible. Like presbycousis nerve deafness due to loud noise tends to affect first the higher frequencies (i.e. above 1000 Hz) but, unlike presbycousis, this loss will often, if severe enough, extend over the whole frequency range.

It would obviously be desirable to establish maximum sound pressure levels that no one should be exposed to for long periods of time if they are to avoid permanent hearing loss. Here we should mention that temporary hearing loss may be caused by noise but if the noise is not too loud and not too prolonged the person will, in time, completely recover his hearing. The period of recovery may be anything from a few seconds to months, depending on the nature of the noise exposure and on the person himself. However, we are concerned here only with permanent hearing loss.

The establishment of maximum permissible sound pressure levels is extremely difficult. This is because, in spite of the vast amount of work done, not enough has been learnt to deal with all the variables. Amongst these variables might be mentioned: the loudness and nature of the noise; whether exposure is intermittent, e.g. a period of exposure and then a period of rest, or continuous throughout the working day; the previous exposure of the person to noise; and the different physical reactions of different people to the same noise. Despite these difficulties a provisional specification of maximum permissible levels has been made (by Rosenblith *et al.*), and is described in Chapter 10. This criterion applies when the person is exposed for a lifetime (at eight hours a day) to the noise; if the specified levels are not exceeded no person should suffer permanent hearing loss.

2

The Behaviour of Sound
in Rooms

We have seen in Chapter 1 that when a sound is originated from a point source in air a series of sound waves proceed outwards from the source in ever-increasing concentric spheres, and that the energy at any point becomes progressively weaker as the distance from the source increases. Eventually, in the absence of any obstruction, the sound becomes so small as to be negligible. Let us now consider in what way this behaviour is modified when the sound source is confined in a room.

The complete behaviour of the sound can be studied most exactly by employing wave theory. This approach is far too complicated for practical use and in architectural acoustics only geometrical and statistical methods are used.

Geometrical Study of Sound in Rooms

The analysis of the paths of the sound waves can be used to find out the path of the direct sound and what kind of distribution the first few reflections of sound will have, or to control these reflections by changing the shape and attitude of the boundary surfaces. In theory it is possible to study the path of the waves during the entire time of their travel over distances of some hundreds of feet. During this time they will have been reflected from the various room surfaces many times, the number depending on the size of the room. In practice the task of plotting the path of the waves beyond the first one or two reflections is too complicated to attempt and in any case what happens to the wave paths after they have been reflected once or twice is not of great importance as will be shown later.

43

REFLECTION OF SOUND

When the sound waves strike one of the room boundaries some of the energy is reflected from the surface, some is absorbed by it and some is transmitted through it. For present purposes the transmitted sound is negligible, and leaving aside also, for the moment, consideration of the absorbed part of the sound, let us consider the nature of the reflection. Suppose that the surface

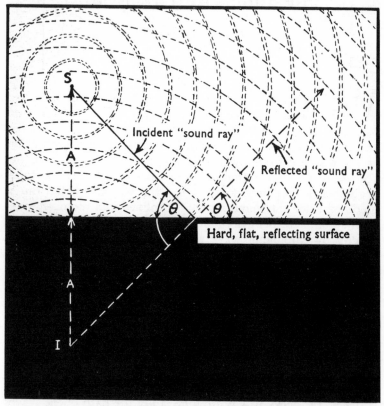

FIG. 10. Reflection from a Plane Surface

encountered is rigid, flat and smooth, we can liken the effect to that which occurs when a light ray is reflected from a mirror. If this comparison is valid we would expect the reflected ray to form as shown in Fig. 10 as though it were produced by a new source (an image I) which is in a position corresponding to the original source but situated behind the surface. This illustrates

44

the familiar law of reflection from which it is seen that the angle of reflection in all cases equals the angle of incidence.

The nature of reflections from curved surfaces can be determined by the simple application of geometrical laws. The very different form of the reflected waves from concave and convex surfaces and the effect of right-angled corners between two flat surfaces are illustrated in Fig. 11.

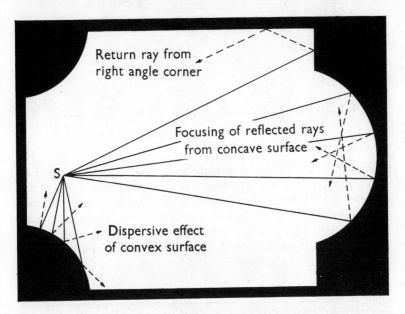

FIG. 11. Reflections from Curved Surfaces and Corners

DIFFRACTION

It is common experience that obstacles do not cast a complete acoustic shadow, as indicated in Fig. 12. This is because of diffraction which causes the sound to be bent round the corner as shown in Fig. 13. This effect occurs in light as well as sound, but its presence in the former case is not so immediately obvious, and can therefore pass unnoticed.

The reason for this is connected with the relative wavelengths of light and sound. Those for light are in the range between 0·00038 and 0·00076 mm, compared with which the obstacles which cast shadows are very large, but the wavelengths of sound are in the range of about 25 mm to 18 m, compared with which the obstacles commonly encountered range from fairly large to

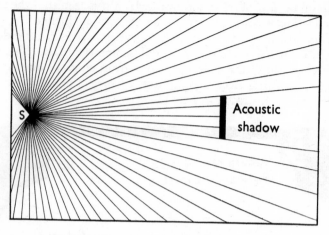

FIG. 12. Formation of an Acoustic Shadow

quite small. It should be remembered that diffraction will be more marked for the low-frequency sounds with long wavelengths than for the short-wavelength high-frequency sounds. For example, consider the effect of a sound wave encountering a 0·6 m wide pier in a room. Sounds of long wavelength, that is much greater than 0·6 m, will diffract round the pier so that there will be practically no acoustic shadow behind it. On the other

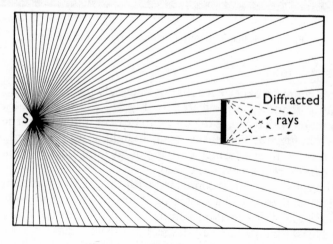

FIG. 13. The Bending of Sound

46

hand, sounds having wavelengths much less than 0·6 m will not be diffracted and the pier will throw an acoustic shadow so that little of this sound will be heard on the far side of the pier.

These circumstances also influence the manner in which reflected waves form, the shorter waves behaving more nearly in the simple geometrical way of light waves as previously described, and the longer waves departing quite widely from this rule.

All the same, the geometrical study of sound paths in rooms is a valuable guide to practical acoustic design, for it must be remembered that the shorter-wavelength sounds play a very important part in the proper audition of both speech and music.

NORMAL MODES

While the sounds with the shorter wavelengths behave like light rays constantly reflected and re-reflected until they finally die away, the longer-wavelength sounds behave in a rather different way. When the wavelength of sound is the same as one of the room dimensions, a 'standing wave' can be set up in this dimension, in which case the sound in the room behaves rather like a pendulum whose rate of swing is determined by the length of the room and whose motion tends to persist for a long time. This same effect also applies when the wavelength is exactly one-half (when two waves fit into it), one-third (when three waves fit), one-fourth, etc., etc., of the room dimension. It also applies to each of the three main room dimensions and to some extent to any subsidiary dimensions between facing flat surfaces (such as the sides of a recess) and even to the diagonals of the room.

For a room of any given size there are thus a number of sound frequencies which receive, as it were, preferential treatment, and if a complex sound containing many different notes is introduced into the room, these particular notes will be accentuated while the sound is present, and may tend to die out more slowly than other notes when the sound ceases. These normal modes, or *eigentones* as they are sometimes called, can be calculated from the parameters of the room sizes. If L, W and H are the length, width and height respectively of a rectangular room, the normal mode frequencies f are obtained from the formula,

$$f = \frac{c}{2}\sqrt{\left(\frac{p}{L}\right)^2 + \left(\frac{q}{W}\right)^2 + \left(\frac{r}{H}\right)^2}$$

where c is the velocity of sound (340 m/sec approx.) and p, q

and r are integers, which specify the mode of vibration. For each mode of vibration (which signifies whether no waves, one wave, two waves, etc., etc., are fitting into the particular room dimension) the integers 0, 1, 2, etc., etc., are substituted for the values p, q and r and all the permutations completed. For example, the first few values (to the nearest Hz) for a room 3 m cube (i.e. L, W and $H = 3$ m) are as follows:

p	q	r	Resonance f	p	q	r	Resonance f
1	0	0	55	3	0	1	170
0	1	0	55	3	1	0	170
0	0	1	55	0	1	3	170
1	1	0	78	0	3	1	170
1	0	1	78	1	0	3	170
0	1	1	78	1	3	0	170
1	1	1	94	2	2	2	183
2	0	0	110	3	0	2	198
0	2	0	110	3	2	0	198
0	0	2	110	0	2	3	198
1	2	1	132	0	3	2	198
1	1	2	132	2	0	3	198
2	1	1	132	2	3	0	198
2	0	2	154	2	1	0	206
2	2	0	154	2	0	1	206
0	2	2	154	1	2	0	206
2	1	2	165	1	0	2	206
2	2	1	165	0	1	2	206
1	2	2	165	0	2	1	206

There is thus a very large number of these normal modes, many of which occur in groups at one frequency. When all the room dimensions are at least as large as the wavelength of the lowest frequency of sound, that is about 10·5 m, the modes lie very close together in frequency, and there is little danger that the sound will be modified noticeably by their presence. In smaller rooms where some of the dimensions are less than 10·5 m it is found that the modes become widely spaced out at the low-frequency end of the range. This is particularly so if two or more of the room dimensions are the same, or are related by simple ratios such as 2:1 or 3:1. The example given above clearly shows this effect. Had the three room dimensions been different from one another many more different frequencies would have appeared in this table.

When widely spaced groups of modes occur sounds can be strongly influenced by them. When for example a single note is sounded at a frequency close to but not exactly the same as one of the strong room modes and it is then cut off, the reverberation will change in pitch. The sound energy has in fact resolved itself into an oscillation at the frequency of the normal mode of the room. If the original sound lies about equidistant in frequency between two room modes then the reverberation may assume a 'vibrato' effect, comprised of the two normal mode frequencies with superimposed fluctuations at the rate of the difference between them.

STATISTICAL STUDY OF SOUND IN ROOMS

In the same way that geometrical study gives information about the arrangement of sound in space, the statistical study of sound is directed towards discovering its arrangement in time.

SOUND ABSORPTION

The nature of the surfaces on which the sound wave falls determines how much will be absorbed. Broadly, hard rigid non-porous surfaces provide the least absorption (or are thus the best reflectors), while soft porous surfaces and those which can vibrate absorb more of the sound. When sound energy is absorbed it is converted into heat energy, although the amount of heat is very small. As the pressure of the air momentarily increases or decreases at the surface of a porous material due to the arrival of the sound waves, air flows into or out of the pores and the friction set up between the molecules of air moving in the restricted space of the pores changes some of the sound energy into heat. Alternatively in the vibrating type of absorbent the surface is set in motion by the alternating air pressure and the friction between the molecules of the vibrating material creates heat.

The efficiency of the absorption process is very simply rated by a number, the ABSORPTION COEFFICIENT of the material, varying between zero and one. If no sound at all is absorbed (an event which never occurs in practice) the sound absorption coefficient would be 0. If all the sound is absorbed, the coefficient is 1·0. Similarly if three-quarters of the sound is absorbed the coefficient is 0·75. Absorption coefficients may be specified as those for sound arriving at all angles of incidence and those for sound arriving only at right angles (normal) to the surface. The first of these coefficients is the one used in architectural acoustic design; it is sometimes referred to as the coefficient obtained by reverberation chamber method of measurement. The coefficient for normal incidence is a more exact scientific property of the material which can in turn be derived from the ACOUSTICAL IMPEDANCE. These quantities enable very accurate computations to be made of the acoustical properties of a room, using wave theory, but the mathematics involved in this approach are far too complex for practical use.

Absorption of any material is not constant at all parts of the frequency scale. Indeed the coefficient for a given material may easily be eight or nine times as great at one part of the scale as at another. Nor is the amount of effective absorption dependent only on the absorption coefficient. It depends to a slight extent on the position of the absorbent material in the room and its relation to other surfaces. For example, if a certain amount of absorbent material is fixed in patches mingled with areas having reflective characteristics they will be slightly more effective than if the same amount of material were fixed all in one area. This is because sound waves arriving at the junction between an absorbing and a reflecting surface are bent in (diffracted) towards the absorbent material. The net effect is that the edges of absorbent material are more efficient than the middle.

The complete picture of the behaviour of a decaying sound in a room therefore comprises a complicated pattern of waves travelling about the room and being reflected at the various surfaces, each reflection reducing the strength of the wave to some extent and probably altering its strength at one part of the frequency scale more than at another. Combined with this motion there may be very long standing waves set up between the various parallel surfaces, particularly if the room is small.

The Reverberation Time Calculation

Each time the sound waves meet the boundary surfaces of the room, some part of their energy is absorbed, while the remainder is represented by the waves reflected from the surface. These reflected waves also eventually meet a boundary and again a part of their energy is absorbed and a part re-reflected and so on. In the lack of continuous replacement of the original sound energy we would therefore expect any sound produced in a room to die away slowly to inaudibility, rather than to cease abruptly when the supplying energy is turned off. How long this dying away process will take must depend on two factors, namely how much absorption occurs when the waves meet the boundaries and how often they do so. As each part of all the boundary surfaces is unlikely to absorb sound equally, the first factor must be discovered from the product of the areas of the different types of surface and their absorption efficiency or coefficient. Then the total absorption is the sum of all these products; the decay time will be less when the absorption is great.

The second factor will depend on the size or volume of the room because sound travels at a fixed speed, and the greater the volume the less often will waves meet absorbing surfaces and the more will be the decay time. This basic relationship was first put into a quantitative form by W. C. Sabine towards the end of the nineteenth century. The now well-known Sabine formula states that the time required for the sound to decay by 60 decibels (that is, the REVERBERATION TIME) is found from the equation

$$\text{R.T.} = \frac{0 \cdot 16 V}{A} \qquad \cdots \cdots \quad (1)$$

where R.T. is the reverberation time in seconds,

> $0 \cdot 16$ is a constant,
> V is the volume of the room (in cubic metres),
> A is the total absorption in [m²] sabins.*

The total absorption (A) is found by multiplying each individual area by its absorption coefficient and adding the whole together —mathematically expressed thus:

$$A = \sum s_1 \alpha_1, s_2 \alpha_2 \ldots s_n \alpha_n$$

where $s_1 \ldots s_n$ are the areas in [m²],

> $\alpha_1 \ldots \alpha_n$ are the absorption coefficients.

The reason for the specification of the reverberation time as being the time required for the sound to decay by the particular amount of 60 dB (or to one-millionth of its initial intensity) is merely to regularise the quantity for reference purposes. The only significance in the choice of this particular amount of sound

* W. C. Sabine called the units "open window units" because they are the equivalent in absorption to a similar area of open window, from which of course no reflection can occur and hence has a coefficient of 1·0. They have since been renamed "sabins" to commemorate his name.

decay is perhaps that it represents the range between a fairly loud sound (say a person speaking in a raised voice at about 1 m distance) and a fairly low background noise at which such a sound would become virtually inaudible. In fact the decay may not take place at a regular rate and be perceived as a linear function at all. Moreover the whole range of the decay is not so important to acoustics as the first part of it. This is more fully explained later.

We can test the equation at its extremities to ensure that it obeys a fundamental law. Suppose we assume a room in which all the surfaces are perfectly reflecting, i.e. have a coefficient of zero. Then the reverberation time will be infinite, and since we have assumed that there will be no absorption of the sound wave in travelling through the air, this is the correct answer. In fact a very small amount of attenuation of the sound wave does take place in its passage through the air, but the amount is quite insignificant for practical purposes except at very high sound frequencies. In very large rooms (where there is a lot of air) this small extra absorption can be added thus:

$$\text{R.T.} = \frac{0 \cdot 16 V}{A + xV} \quad . \quad . \quad . \quad . \quad (2)$$

where all the quantities used in formula (1) are the same in this formula, and x is the air absorption coefficient. This is slightly dependent on the temperature and humidity as well as sound frequency, but for practical purposes these variations are too small to matter; values are given on p. 312. The value of $0 \cdot 02$ at 4000 Hz shows that the amount of absorption will be quite small except in rooms of large volume.

If we now test the Sabine formula at the other extremity, and assume a room having all its surfaces perfectly absorbing (coefficient $1 \cdot 0$) we then obtain an anomalous result. For example, assume a room of 3 m cube, then $V = 27$, $S = 54$, $\alpha = 1 \cdot 0$.

$$\text{R.T.} = \frac{0 \cdot 16 \times 27}{54 \times 1 \cdot 0}$$

$$= 0 \cdot 08 \text{ sec}$$

But reverberation time should be zero, because no sound is reflected from the boundaries. For practical design this discrepancy is immaterial except in rooms where the total absorption is very great (i.e. average α approaches $1 \cdot 0$), but this circumstance applies only to such rooms as broadcasting or sound studios and to certain very 'dead' rooms used for scientific measurement purposes. For the great majority of cases the Sabine formula is sufficiently accurate, and is accepted as the standard method of calculation, for instance, in examinations set for architectural qualification. A worked-out example of the use of the formula is given in Chapter 3.

Two or three proposals have been made for the modification of the Sabine formula to overcome its anomalies. Of these the modification due to Eyring is the one which is most often used. This is in the form

$$\text{R.T.} = \frac{0 \cdot 16V}{S[-\log_e (1 - \bar{\alpha})] + (xV)} \qquad . \quad . \quad (3)$$

The term xV is the air absorption and can be omitted except for very large rooms and high frequencies as explained above. The

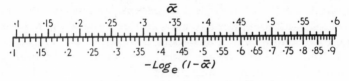

Fig. 14. Values of $\bar{\alpha}$ against $-\log_e(1 - \bar{\alpha})$

term $\bar{\alpha}$ is the average absorption coefficient of all the surfaces of the room and is obtained by working out the total absorption as described above and dividing by the total surface area, thus:

$$\bar{\alpha} = \frac{\sum s_1 \alpha_1, s_2 \alpha_2 \dots s_n \alpha_n}{S}$$

$\log_e$ is the Napierian or hyperbolic logarithm (actually 2·3 times the common logarithm to the base 10), but the values for $-\log_e(1 - \bar{\alpha})$ for each value of $\bar{\alpha}$ are given in Fig. 14. As we have said above, there is no need to use this formula in preference to the Sabine formula except when $\bar{\alpha}$ exceeds about 0·25. A worked-out example is given in Chapter 5, Table IV.

All reverberation formulae embody the assumption that the sound energy is evenly spread over all the surface area of the room and that the sound is perfectly diffuse. In practice this is rarely absolutely true. We have seen, for example, that standing waves can be set up in a room particularly at the lower frequencies, and when this occurs the distribution of sound pressure on the surface varies quite widely. One can detect the presence of these irregularities quite easily when a note of a single frequency such as the 440 Hz tuning note which precedes the opening of certain B.B.C. broadcasts is heard, by closing one ear and moving about the room. This is why the predictions of reverberation time particularly at the low frequencies may often be inaccurate. Nevertheless, the calculation of this quantity remains an invaluable tool in the acoustical design of rooms.

Effects of Reverberation on Subjective Acoustics

If the boundary surfaces of a room are highly reflective (the reverberation is long) when a sound of certain energy is introduced, then because of the constant reflection of this sound we would expect the loudness to be greater than that if the same sound were made in open air where no reflections occur. Similarly if more and more sound absorption is brought into the room we would expect the sound level to get less and less, until if all the surfaces could be made perfectly absorbing the level would fall to the same value as would be obtained in open air. The total sound level in any room is therefore made up of two parts, that produced by the direct path of the waves from the sound source to the listener and that produced by reflections from the walls and other surfaces.

These two parts are known as the direct sound and the reverberant sound. The first part depends only on the strength of the source, but the second part depends also on the amount of absorption in the room. If our object is to obtain the loudest possible sound, it would seem desirable to make the reverberation as long as possible, that is, to reduce the amount of absorption to a practical minimum, but before accepting this generalisation we must examine what effect reverberation has on the distribution of the sound energy in time.

When a steady sound is started in a room the volume level does not rise instantaneously to its final value, but builds up over a certain time. The increase in sound is at first very rapid until a level is reached which is due to the direct path of sound from the source. After this a further increase takes place but at a diminishing rate due to the build-up of the reverberant sound in the room. As the direct sound level is often only a little less than the total and final sound level the subjective effect is that sound onset is almost instantaneous, the slight and slower stages of build-up often passing unnoticed. However, as we have already seen above, when the sound source is turned off an appreciable and usually noticeable time elapses (again depending on the reverberation) before the level falls to zero. This means that the hearing of transient sounds (i.e. those having sudden starting or stopping characteristics) will be somewhat modified by the presence of reverberation. Speech and music consist very largely of transient sounds and if we are to preserve intelligibility of speech and natural quality of music these transients must not be seriously affected. The transients will, of course, be modified to some extent in any practical room, but provided the modification does not exceed certain limits, all will be well. Otherwise it would be impossible to communicate by speech or listen with appreciation to music except in the open air.

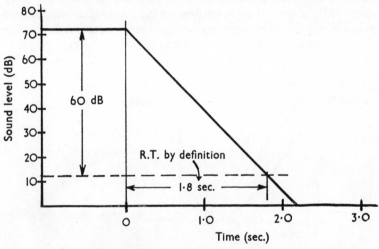

Fig. 15. Idealised Decay Curve

Briefly stated, it is necessary that the reverberant energy from one transient sound shall have fallen to a low enough value not to obscure or mask the appreciation of the following transient. Now speech transients, generally syllables or consonant sounds, follow one another at a rate of about 10 to 15 per second and music sounds at a rate which depends on the music but can be as high as 20 notes a second. It is therefore necessary that reverberation shall not be so long nor have a character which will cause these quickly following sounds to become blurred together.

We have assumed up to now that a sound will decay regularly, as shown in Fig. 15, but in practice we find that the decay may show irregularities such as those in the actual examples given in Fig. 16. It is plain that if these irregularities are too large, and particularly if they take the form of the curve illustrating the result of an echo, the chances of properly distinguishing the rapidly following transients are greatly reduced. The double slope decay curve is a type frequently found where two spaces (or rooms) are joined by an opening, and one of them has a much longer reverberation than the other. Alternatively, it may result from a normal mode which is insufficiently damped, the first steep part of the curve representing the average rate of decay, and the second part the rate of decay of the mode. In either case this type of curve can lead to undesirable results. Because the absorption coefficients of different materials vary quite widely with the sound frequency, the reverberation time is liable to vary. If extreme variations are permitted it is possible that the long reverberation (and the loudness) from a sound of

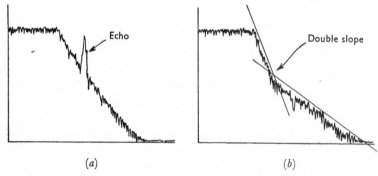

(a) (b)

FIG. 16. Actual Decay Curves, showing (a) an Echo and (b) Double Slope

one frequency will obscure the perception of the following sound which may well be of a different frequency, loudness and reverberation.

For the foregoing reasons it is the general aim in acoustic design to provide a certain amount of reverberation (an optimum reverberation time which experience has shown will depend on the size of the room), to ensure that this value is roughly constant over the whole audible range of frequencies and that each decay takes place at a reasonably steady rate without excessive fluctuations. Lastly, everything must be done to allow the direct sound which will not be modified by the room acoustics to reach every listener at the greatest possible strength.

CONTROL OF SOUND DISTRIBUTION BY ROOM GEOMETRY

In small rooms the direct sound will generally be sufficiently loud at any point in the room to ensure good hearing conditions provided that the reverberation time is within certain limits. In large rooms, however, the direct sound will have fallen to a low level at points distant from the source and it may be necessary to supplement it by arranging for strong reflections to be directed towards that part of the audience which is most distant from the sound source.

Splayed, flat or slightly convex surfaces can be used in plan and section to direct reflections towards the seating area, the angles being determined by the geometrical laws of reflection outlined above. Such reflections are particularly valuable if they follow very shortly in time after the arrival of the direct sound, which is the same thing as saying, which arrive at a position having travelled a minimum distance further than the direct sound (see also Fig. 25, p. 71).

Conversely, reflections which arrive at a point having travelled more than 15·5 to 18·5 m further than the direct distance to this point may give rise to echoes. This acoustical defect is most likely to occur when concave surfaces focus sound rays subtending a large arc at the source on to a point. For example, large domes or barrel vaults should generally be avoided for this reason.

The focusing of sound from a concave surface will occur only if the approximate diameter of the curvature is at least as large as the wavelength of the sound. For example a coved cornice section of 1·2 m diameter will not focus any sound of lower frequency than about 250 Hz. A curved rear wall of, say, 46 m diameter curvature will on the other hand focus sounds down to the lowest frequency, *provided* it is at least approximately a wavelength in height—say 6 m or so. If one dimension is much less than this—for example, the piece of wall visible above the heads of the last row of audience is sometimes only about 1·5 m high— then it is unlikely to focus sounds at the very low frequencies, although it will of course focus all sounds of 1·5 m wavelength and less (200 Hz and greater).

This focusing of sound is a major fault if the geometry of the room is such that the focus point is anywhere near the audience. If it is out in the room space, i.e. well away from any position in which audience can sit, it is certainly less harmful and may not matter at all. Apart from changing the major shape of the surface so as to remove the focusing effect there are two methods which can be used to prevent focusing. These are both frequency dependent and must be designed accordingly. Broadly, we can either absorb the sound or diffuse it. Absorption we have already discussed, and obviously if it is to be effective the absorbents used must have a high coefficient (say exceeding 0·4) at all the frequencies the curve would focus.

Diffusion can be obtained in two ways, either by modelling the surface with projections or by changing its absorbing nature, i.e. by having patches of absorbent and reflecting materials mixed together. Projections will be effective only if their dimensions are large enough. The minimum dimension required is at least one-tenth of a wavelength, therefore for effective diffusion down to the lowest frequencies the projections should not be less than. 0·75 m. Various shapes of the projections are possible, but convex shapes are particularly good because reflected waves are widely dispersed, as has been shown.

Alternatively it may be possible to change the angle of the curved surface so that the focus does not come within the room at all. Suppose, for example, the curved rear wall mentioned above were canted inwards at the top, then the sound reflections would be directed downwards on to the rear rows of audience, helping to reinforce direct sound, and the focus could be below the level of the auditorium floor, outside the confines of the room.

Another type of echo sometimes occurs between pairs of plain parallel surfaces. This is known as flutter echo and can often be observed if an impulsive sound such as a hand clap is produced in a room where one pair of parallel surfaces is hard and non-absorbing and the other pairs of surfaces are relatively absorbent or diffusing. It occurs for example in a rectangular room where the floor is uncarpeted, the ceiling flat and hard but the walls are provided with windows, curtains, recesses, furniture, etc., all of which absorb or diffuse the sound. This defect can be avoided by setting the surfaces out of parallel by a few degrees, but this solution sometimes brings about other disadvantages. Generally it is better to apply diffusive or absorptive material to one or both of the surfaces causing the echo. Right-angled corners which break up the reflections are useful for diffusion, particularly when there are many of them, as for example when they are formed by breaking up the surface into two or more planes as in coffered ceilings.

CHARACTERISTICS OF SOUND ABSORBENTS

Every surface of a room, of whatever material, and all the objects in it, whether animate or inanimate, will absorb sound in some degree. Many of the usual surfaces, such as plastered brick walls, windows or woodblock floors, do not absorb sound very well. In practice the actual absorption characteristics of many of these common materials depend on the way in which they are used. For example, the plastered surface of an 0·45 m brick wall will have a rather different absorption characteristic from a plasterboard ceiling nailed to joists. Also there are very numerous slight or major variations in building techniques and often no data are available either because no one has troubled to measure the coefficients or more often because it is impractical to do so.

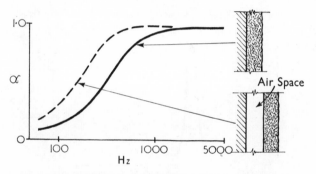

FIG. 17. Absorption Characteristics of Porous Absorbents

A list of approximate absorption coefficients for various common building finishes is given in Appendix A. Sound absorption is not an intrinsic property of a material alone and for this reason the values given should be regarded as representative rather than precise.

Turning now to materials having high absorption, that is, those materials which we may introduce intentionally into a room to correct or modify its reverberation characteristics, it is clear that a complete specification of absorption coefficient can only be expressed in the form of a graph (see Figs. 17, 18 and 19). The first of these (17) shows the performance of a typical porous absorbent, 25 mm rock wool mounted directly against a solid hard surface. This shape of absorption characteristic is common to the great majority of proprietary acoustic treatment materials, such as acoustic tiles, and to many of the fortuitous sound absorbers, such as curtains, furniture, carpets and people. Its main feature is the considerable reduction in effectiveness at the low frequencies. There is a direct relationship between the wavelength of the sound and the thickness of the porous material for maximum absorption. Thus

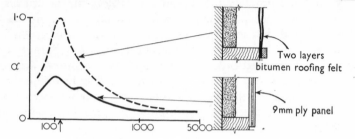

FIG. 18. Absorption Characteristics of Panel Absorbents

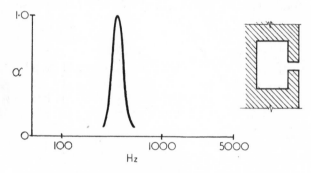

FIG. 19. Absorption Characteristics of a single
Helmholtz Resonator

thin materials can effectively absorb only those sounds having short wavelengths. Some improvement in the lower-frequency absorption of a thin material can be obtained by spacing it away from the solid backing, and the dotted lines on Fig. 17 show the changes produced by spacing the rock wool away from its backing. A similar change in the curve is obtained when the material is made thicker.

In assessing the probable coefficients of porous absorbents the percentage volume porosity, surface porosity and degree of interconnexion between the pores are important factors. These properties influence the flow resistance, i.e. the degree of difficulty with which air flows into and out of the material, and

hence the absorption. Materials in which the pores are not connected, such as some of the foamed resins and cellular rubbers,
cannot have very high absorption. If a porous absorbent is
decorated with a film-forming paint on the surface and many or
all of the pores are closed, a considerable loss of absorption will
occur. This is the main disadvantage of the use of such materials
as acoustic plasters. To overcome this difficulty many proprietary materials such as soft wood fibreboards and certain
mineral wool and plaster products are prepared with the surface
drilled, grooved or fissured with a quantity of holes. The purpose
of this is to expose a porous surface (the sides and bottom of the
holes) which cannot easily be filled with paint. The flat external
surface may then be finished in the factory with a surface coating
which is more or less non-porous, and subsequent painting,
provided that it does not fill the holes, makes no difference to
the absorption. A warning must be given that a number of perforated materials are not surface coated and a marked change
in performance must be expected when eventually these
materials are decorated for maintenance purposes.

The second type of absorbent is the panel or membrane
absorber having a characteristic curve as shown in Fig. 18, which
shows the coefficients for a 10 mm thick wood panel mounted
over a 38 mm air space and in front of a hard backing. Any
plate or layer of non-porous material mounted with an air
space between it and a solid backing will operate in some degree
as a panel absorbent. The main feature of the characteristic is
a high coefficient somewhere in the low-frequency part of the
sound range and a falling absorption at frequencies above this.
The position of maximum absorption depends on the resonant
frequency of the panel, which can be likened to any other
mechanical system having resonance, e.g. a weight supported
on a spring. In the analogy the weight is represented by the mass
of the panel and the spring by the compliance of the air enclosed
behind it and to some extent by the elasticity of the panel
material. The resonant frequency of panels of practical weights
and spacings falls within a range of 40 to about 400 Hz, and is
calculated from the formula

$$f_{res} = \frac{600}{\sqrt{md}}$$

where m is the mass of the panel in Kg/m² and d is depth of the
air space behind it in centimetres. This relationship is displayed
in a convenient graphical form in Fig. 20.

Above this frequency the necessary requirement that the dimensions of the panel shall be small compared with the wavelength of the sound is unlikely to be met. Apart from this requirement the size of the panel makes little difference to its absorption characteristics.

Measurements show that the position of maximum absorption can be affected, in stiff panels, by the elasticity of the panel

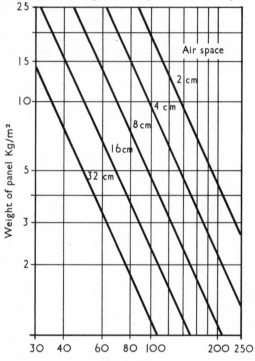

FIG. 20. Resonance Frequency of Panels

material and nature of the edge fixing. For example, the calculated resonant frequency of the plywood panels shown in Fig. 18 is indicated by the small arrow. There may also be secondary peaks of absorption in the curve. However, the deviation is rarely of practical significance. When the material used is one having little or no elasticity, such as bitumen roofing felt, then a practical curve closely approaching the ideal one is obtained. The actual measured absorption for a bitumen roofing felt panel with the same resonance frequency is shown dotted in Fig. 18. Here the resonant peak agrees exactly with the calculated frequency. The absorption coefficient is often increased and extended over a broader range in frequency if a porous material such as glass or rock wool is inserted in the cavity behind the panel. This introduces increased damping by resisting the free movement of the air in the cavity when the panel vibrates.

The membrane type of absorber has application where objection is raised to the holes in many acoustic materials on the score of hygiene. Employing a fairly light-weight, supple membrane such as leather cloth over a low-density porous material such as glass wool will enable the resonance to be raised to some frequency in the middle range, say 250 to 500 Hz, but it is doubtful if it is possible to obtain high coefficients in the higher-frequency range (1000 Hz and upwards) where much sound energy occurs in noise.

The last class of absorbent is the cavity or Helmholtz resonator. This can take many forms, although all basically consist of an enclosed body of air which is connected by a narrow passage or neck with the space containing the sound waves. That an empty bottle has the properties of a cavity resonator can readily be demonstrated by blowing across the neck. This demonstration also illustrates that the resonant effect is sharply confined to one pitch of sound. The absorption provided by a single resonator depends on a great many factors, but it can be said that in general a high efficiency of absorption is obtained at the resonant frequency but that absorption is very little for all other frequencies except those very close to the resonance point (see Fig. 19). Single cavity resonators can be designed to provide absorption at any point in the frequency scale, but owing to their sharp tuning they are not often used for general acoustic treatment where a major change is required, but only where a particularly long reverberation is experienced at a single frequency such as that due to a normal mode, and it is desired to reduce this without greatly affecting the average reverberation.

A much commoner application of the cavity resonator principle for sound absorption is a material consisting of an airtight plate or panel drilled or punched with a pattern of holes or slots (which form the necks of the resonators) mounted with an air space between it and a solid backing. The portion of this air space behind each hole forms the body of the resonator and in most cases there is no need to divide off the separate resonator bodies by partitions (see Fig. 21).

As in single resonators the dimensions of the neck and body

determine the resonant frequency but the multiple resonator is not so selective in its absorption. The usual type of absorption coefficient is given in Fig. 22, which is for 20% slotted hardboard, 3 mm thick, mounted at 50 mm from a hard backing and with glass wool or porous paper in the cavity as 'damping'. The calculated resonance point is about 1200 Hz and it will be

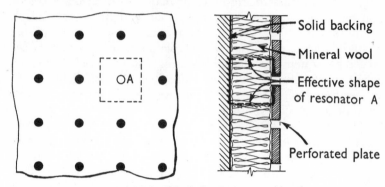

FIG. 21. Multiple Helmholtz Resonator Absorbent

seen that this agrees fairly well, but not exactly, with the position of maximum absorption. This shift in the resonance point is due to the presence of the glass wool and the amount of the shift will vary with the density and properties of the porous

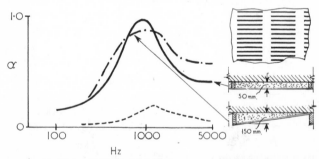

FIG. 22. Absorption Characteristics of a Multiple
Helmholtz Resonator Absorbent

material used in the cavity. It is important to insert this 'damping' material because it has two useful effects. First it increases the effectiveness of the absorption at resonance point (the curve for the same multiple resonator without any damping material is shown dotted in Fig. 22) and secondly it produces extra absorption at frequencies above the resonance, particularly when the holes total a substantial proportion of the whole panel area.

Taking the following practical values for the various dimensions,

Thickness of plate	1·5 to 12 mm
Diameter of holes	1·5 to 7·5 mm
Spacing of holes	12 to 100 mm
Air space behind plate	6 to 100 mm

the range of frequencies at which resonance occurs is from about 90 Hz to 4000 Hz. Thick plates with large-diameter holes, well spaced and with a large air space behind, will give the lowest frequencies. Changing any of the dimensions in the direction of the other extreme will raise the resonant frequency.

Maximum values of absorption coefficient for this type of absorber are between 0·6 and 0·9 and the coefficient is often at least one-half of the maximum over a range of at least three octaves when there is a porous material behind to provide damping. Most of the proprietary punched, drilled or slotted boards at present available have regularly spaced holes. This means that each resonator has the same frequency. If the spacing or size of the holes were varied it would be possible to produce a material having resonators tuned to different frequencies and thereby to obtain a coefficient curve flat, or to any desired shape. Alternatively, but perhaps not quite so effectively, the variation in resonance can be achieved by varying the space behind the panel. A curve for the coefficient of the 20% slotted panel mounted so that the air space varies between 6 mm and 76 mm is given dashed and dotted in Fig. 22. The physical form of multiple-resonator treatments is thus capable of development into a great number of types.

When absorbing materials are used to form a shaped unit, such as a sphere, double cone, cylinder or cube, the result is called a 'functional absorber'. These objects are suspended freely in a room space, some distance from the boundary surfaces, and because the sound waves in their vicinity are diffracted towards them, they provide a powerful absorbing effect, and can be used when the boundary surfaces are not available for the application of normal absorbents. Maximum absorption over a wide frequency range is obtained when their greatest dimension is between 0·45 and 0·9 m and the acoustic impedance of the surfaces is within certain limits.

3

Design of Rooms for Speech

The main aim in designing rooms for speech is to ensure that every member of the audience can clearly hear what the speaker says; in other words, the problem is essentially one of intelligibility. A secondary aim is to preserve the natural qualities of a speaker's voice to ensure that each audience member can appreciate the nuances and dramatic effect being sought by the speaker. This requirement applies mostly to the theatre of course, and perhaps hardly at all to the public meeting hall.

Nature of Speech Sounds and the Effects of Room Acoustics

Speech sounds consist of a flow of various combinations of vowel and consonant sounds. These combinations are woven into a main structure consisting of certain predominant tones which are natural attributes of the person speaking. These voice tones (sometimes called formants) can be varied over a small pitch range by the speaker, to give emphasis or shades of meaning, but are always to some extent distinctive of the person. For example, the major difference between male and female speech is the pitch region in which the formants occur. The formants give the basic tone to speech and are heard most in the vowel sounds. The consonant sounds are nearly all of a transient nature; that is to say, they are very short and rapidly changing sounds, very often 'unvoiced' (i.e. containing no formant tone) and therefore having very little acoustic power. It is the correct recognition of consonant sounds which is the principal factor in speech intelligibility in an auditorium.

If every member of an audience is to hear well it is obvious that the sound at each audience seat must be loud enough. A speaker can raise his voice to adjust its loudness to suit the size of audience he is addressing, but there is a very definite limit to his capabilities. This assumes that he is relying on the unaided power of the voice and will not be using speech amplification, which will be discussed later. It is therefore essential that the best use be made of the limited amount of acoustic power available, especially if the hall is large, and that there shall be an absolute minimum of background noise such as that from traffic outside the building or ventilation plant, which will tend to mask the wanted sound. We have also seen that reverberation will raise the loudness above the level obtained in its absence; it is common experience that it is more difficult for speech to be well heard at any great distance in very dead acoustic conditions, such as are found in the open air, than in rooms with good acoustics.

Although reverberation will add to the loudness it is well known that too much reverberation can be very harmful to good hearing. In some large buildings, such as cathedrals for example, reverberation may be as long as 14 seconds. Now when speech is heard in such a building, two effects are immediately apparent. First intelligibility suffers very severely because the sound of each syllable is obscured by the still present reverberations of previous syllables. Secondly, although the speaker may have no great difficulty in providing sufficient sound energy, he may be inhibited from speaking as loudly as he could because of the peals of reverberation set up by every sound he utters.

Intelligibility tests have been made in rooms of different size and having different amounts of reverberation, and the results show (Fig. 23) that percentage syllable articulation progressively decreases as the reverberation time and the volume of the room increase. In practice a speaker may naturally tend to speak more slowly and loudly when in a large room addressing a large audience, and this will to some extent help to counteract the decreasing intelligibility caused by the reverberation.

BASIC ACOUSTIC REQUIREMENTS OF LARGE ROOMS FOR SPEECH

It is important that the direct sound, i.e. sound proceeding directly from the source to the listener without having been reflected from any room surface, shall be as strong as possible. We have seen that this sound weakens with distance according to an inverse square law, and therefore the average distance between the source and the listener should be kept as small as possible. It is also important that this direct sound path shall not be obstructed. This not only means that there should be no part of the building such as columns or balcony fronts interposed in the path, but also, because sound is absorbed very strongly when it passes at grazing incidence over an audience, that the seats should be arranged so that the heads of one row of the audience

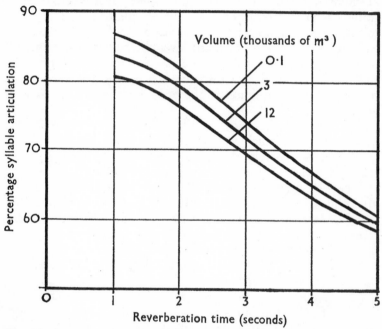

FIG. 23. Average Percentage Syllable Articulation against Reverberation Time in Auditoria of various Volumes

do not obstruct the direct sound paths to the people in the row behind. The best way of ensuring this is by seating the audience so that a clearance of at least 76 mm is provided between the sight line from one row and the sight line from the next. The clearance can, with advantage, be 100 mm or more. This principle is best shown in the diagram Fig. 24, from which it can be seen that a curving rake is the ideal shape. This can be simplified into a straight rake, but this should not be done so that the clearance at any part of the seating is reduced below the desirable minimum. This arrangement also provides, of course, a good view for the audience. It is axiomatic that if the audience cannot see the speaker well, there is little chance that they will hear him well. It is frequently necessary, for other reasons, to use a flat floor to the auditorium. Then the best that can be done is to

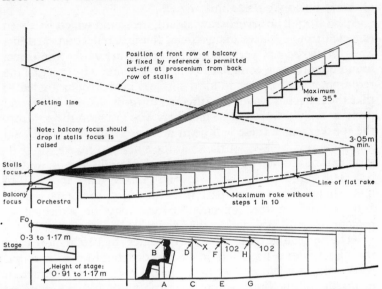

Position of front row of balcony is fixed by reference to permitted cut-off at proscenium from back row of stalls

Maximum rake 35°

Setting line

Note: balcony focus should drop if stalls focus is raised

3·05m min.

Stalls focus

Balcony focus · Orchestra

Line of flat rake

Maximum rake without steps 1 in 10

Fo

0·3 to 1·17 m

Stage

B | D | X | 102 | H | 102
F
A | C | E | G

Height of stage: 0·91 to 1·17 m

The method of setting out, as demonstrated by the lower diagram and the following notes, is based on the assumption of 1·12m as the height from the floor to the spectator's eye level and 102 as the distance from the eye to the top of the head.

Method of setting out:

Draw level of stage and front row of seats (stage: 0·91 to 1·17 m above level of latter)

Draw perpendiculars to represent seat rows

LACHEZ'S SYSTEM

Fix point Fo on setting line (0·3 to 0·91 m above stage)

Mark off AB = 1·22 m and draw FoD through B cutting CD at D

Mark off DX = 102 and draw FoF through X cutting EF at F, repeat process for each row

The points thus found represent the eye level for each row, by measuring 1·12 m down on the perpendicular from each of these, the points obtained mark the curve of the floor

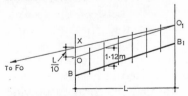

O_1

X

B_1

$\frac{L}{10}$ O 1·12m

To Fo | $\frac{L}{10}$ | B

L

Find eye level O of first row (as described above)

Draw L = depth of balcony front to back

Fix point X vertically above O so that $OX = \frac{L}{10}$

Draw FoX and produce to cut back line of balcony

Draw OO_1 the floor line BB_1 will be parallel to OO_1 at a vertical distance of 1·12 m below it.

SIMPLIFIED METHOD FOR BALCONIES

THEATRE SEATING DESIGN OF RAKED SEATING

FIG. 24. Setting out of Raked Seating

raise the speakers' platform sufficiently high to ensure that minimum clearance is obtained at the rear rows of the hall.

In plan, seats should be ranged so that none falls outside an angle of about 140° subtended at the position of the speaker. This is because speech is directional, and the power of the higher frequencies on which intelligibility largely depends falls off fairly rapidly outside this angle.

Apart from true reverberant sound, i.e. sound which has been reflected from surfaces many times, the first reflections arriving at a listener's ears very shortly after the arrival of the direct sound operate as a contribution to direct sound. In practice the requirement is to provide hard surfaces in the room angled so that the reflected waves are directed towards the audience, and preferentially towards those members who are most distant from the speaker, since these will be in most need of some additional sound energy. Splayed surfaces at the sides and above the platform can be studied geometrically in plans and sections, and all reflected sound paths which are no greater than about 10 m more than a direct sound path from the source to the part of the audience where the reflection arrives may be expected to provide a useful addition to the 'direct' sound.

Reflected sound paths which are greater than 15 m more than the direct sound path should be avoided because such sound waves can result in echoes being heard and a great reduction in intelligibility. Concave curved surfaces should generally be avoided, particularly if their geometry is such as to cause reflected waves having long path differences to be focused on part of the audience. Two sections of a hall are given in Fig. 25, in the first of which are shown 'useful' reflections from a correctly designed ceiling, and in the second a bad ceiling shape which would be liable to cause a serious echo to develop. The side walls of a room may be parallel but are probably better converging on to the platform (giving a fan-shaped plan), because this reduces the average distances between the speaker and the audience. If galleries are required the free height between the heads of the audience and the gallery soffit should be made as great as possible, and the depth of the seating area under the gallery should not exceed twice this height.

If all the surfaces of the room are planned so as to provide useful reflections, that is, there is none which, because of its position or shape, is liable to cause echoes, then there is no reason why they should not all be of highly sound-reflecting materials. This would mean that the only highly sound-absorbing surface in the room would be the audience and their seats.

We have seen that the reverberation time must not exceed

certain limits, and that it is proportional to room volume and inversely proportional to the amount of absorption. There is consequently a certain approximate volume of room space which when it contains the absorption provided by one mem-

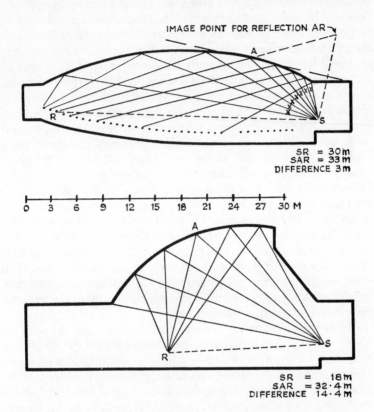

FIG. 25. Two Long Sections of Halls, (a) with usefully reflecting Ceiling, (b) with Ceiling which will cause Echoes

The ceiling shape in the top illustration is designed as follows: It is assumed that roughly an equal amount of sound energy is radiated in each of the sectors where the angles x are equal. The ceiling is shaped so that the energy is more concentrated on to the audience as the distance from the sound source increases. For example, the energy in the second sector falls on about one-half of the amount of audience compared with that on which the first sector energy falls, and so on.

ber of audience will have a reverberation time of about the right amount. Allowing for the fact that the room surfaces, although nominally reflecting, will have some slight absorption, this volume comes out at about 2·8 m³ per seat. If the above condition can be met (i.e. all surfaces useful reflectors), then it is best to design the room volume to this value, because in this way the average sound energy reaching the audience will be at a maximum. If the volume is made greater than this (and it may be considered necessary, to give good architectural proportions), then additional absorbents other than the audience will have to be introduced to achieve an acceptable reverberation time. It is therefore best to keep the volume to as near to 2·8 m³ per seat as possible, and it is strongly recommended not to exceed 4·2 m³ per seat, particularly in rooms for very large audiences.

There will obviously be a maximum size of room in which it would be reasonable to expect adequate hearing of unamplified speech. This size will depend very much on how low the ambient noise can be kept, and on the correct design of the room shape and surfaces. No exact value for this maximum can be given except perhaps to say that it is probably less than 8500 m³. However, in rooms of considerably less volume than this it is common to find speech-amplifying systems in use. There are two main reasons for this, one being that the art of clear public speaking appears to be waning, probably largely owing to the ubiquitous microphone, and the other that few public halls have been properly designed in all the respects we are discussing. If it is known from the inception that sound amplification will always, or almost invariably, be used in the room, then the acoustic requirements are much simplified.

Having decided on the general planning, size and shape of the room, the next step is to design the surfaces so that the reverberation time will be suitable. The graph Fig. 26 (after Knudsen and Harris) shows a suggested relationship between average reverberation time with maximum audience and volume of hall. In practice a time of between 0·75 and 1·0 second is suitable in most rooms. There is no great objection to having slightly less reverberation than is indicated, but at the same time the reverberation must not be very much less or the speaker will have difficulty in filling the hall with sound and an unnaturally 'dead' condition will result. In calculating the reverberation time (a detailed example of this calculation is given later) it will be noted that the audience themselves provide a very large proportion of the total sound absorption, particularly in a large hall. This means that in the absence of an audience, or when there are many empty seats, there will be a marked change in the reverberation time. This difficulty can be partially overcome by using seats which provide almost as much absorption when they are empty as when occupied. Only well-upholstered theatre-type seats will give this amount of absorption, and if other considerations forbid the use of such seats it is better to design the room for the optimum reverberation time with only, say, one-half or two-thirds of the audience present. Any type of seat, however hard, provides some absorption of course. Approximate values are given in the table in Appendix A.

Having decided what amount of sound-absorbent treatment is required, it remains to consider where this material should be located and what form it should take. Broadly, it is best to place the greater part of the sound absorbents at the end of the room

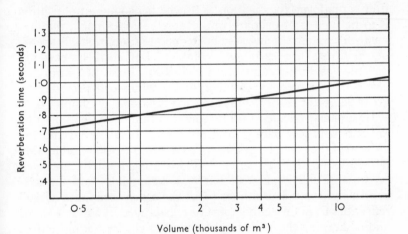

Fig. 26. Recommended Reverberation Times for Speech (after Knudsen and Harris)

away from the speaker, and to apply them to those surfaces which might produce unwanted echoes, for example the rear wall of the hall, particularly if the risk of making this a concave curve has been incurred, and the gallery front.

Absorbent materials of a hard-wearing kind, such as perforated boards or wood strips over glass or rock wool, should be used on walls where they are liable to suffer damage or wear. The less rugged materials, such as fibre acoustic tiles, should be used only out of reach of the audience.

We have dealt at length with the design of a large room for speech audition from a platform and in doing so have enumerated all the basic design problems. There are other kinds of speech rooms, in designing which all or most of the basic design requirements must be observed. There are, however,

73

particular problems relating to some of these rooms which are described below.

ROOMS FOR DEBATE

Most of the foregoing remarks have assumed that there is only one position for the speaker (this may be taken to include one area, such as a platform and not merely a single point). The requirement of good audition from *any* point in the room arises in debating chambers. This makes the planning more difficult in that it is usually impossible to design reflecting 'splays' which are equally effective for sources of sound anywhere in the seating area. In debating rooms, even more than in rooms for address from a platform, the decision as to whether sound amplification is to be used is of primary importance. The technical requirements for an amplifying system to operate from multiple speaking positions are extremely complex, but such systems are possible, as for example in the British House of Commons, although it is very doubtful if such systems can provide any great degree of reality. This aspect is discussed in Chapter 6.

If it is early decided that no sound amplification will be used an extremely low background noise must be achieved and the volume of the room must be kept as small as possible, preferably not exceeding $2 \cdot 8$ m^3 per seat. The seating must be arranged on suitable steep rakes so that *each* person has good uninterrupted sight and sound lines to *all* the other occupants. Useful reflections from the ceiling must be encouraged by suitable shape and by keeping the ceiling level not too high, i.e. not more than $7 \cdot 6$ m.

Frequently debating chambers are provided with public seating space. Although these seated in such areas need to hear the debate well, there is no need to ensure good hearing of sound from the public space into the actual debating area, in fact rather the reverse. The public area may take the form of one or more galleries which may be at some distance from the debating floor. It is desirable that these parts of the room, particularly if they are in the form of adjuncts, should be treated as acoustically 'dead' areas by the use of carpet (which will prevent unwanted foot traffic noise as well as absorbing air-borne sound), upholstered seats and sound-absorbent ceiling or walls or both.

MULTI-PURPOSE ROOMS

In many cases an auditorium is required to be suitable for both speech and music of one form or another. In this class are included churches, theatres, school assembly halls, community halls, etc. Although the ideal acoustical conditions for music are not exactly the same as those for speech, for rooms of up to 2800 m³ in volume there is very little practical difference. For larger rooms the choice of a suitable reverberation time must be a compromise. The primary purpose of the room and also tradition must influence the choice. For example, the assembly hall for a large comprehensive school will probably be used for theatrical performances, music and the showing of sound films, but we might concede that the primary purpose is for speech, addressed to the scholars. It is best, therefore, to aim for the ideal conditions for speech intelligibility and accept the fact that music heard in this hall will suffer a little from lack of 'tone'. On the other hand, a church with a tradition of choral singing accompanied by an organ demands a different compromise, in fact one favouring musical acoustics. Recommendations for this type of building are given in greater detail in Chapter 4.

If the room is to be provided with a sound-amplifying system then there is less need to attempt a compromise for the best acoustical conditions; the reverberation time can be made long enough for music and the sound-amplifying system used for speech can be designed to overcome the comparatively long reverberation. The design of such systems is discussed in Chapter 6.

Two electronic systems are available which are claimed to increase the subjective reverberation time of a hall, and so can be used to adjust the acoustic conditions to suit the programme material. These are ambiophony and assisted resonance (see page 101).

School halls with reflecting flat ceilings and floors without fixed seating are sometimes found to be acoustically bad for speech when only a small number of audience is present. This is due to flutter echoes arising between floor and ceiling, see p. 88.

THEATRES

All the general recommendations given above apply to the acoustic design of theatres, and in particular it is of great importance to arrange the audience compactly so that the average distance from seating to stage is as small as possible. Good sight lines are, of course, absolutely essential.

Satisfactory amplification of stage performance is, at present, largely an unsolved problem (see p. 139). While actors have to rely on the unaided power of their voices, theatres seating more than about 1500 persons are likely to be unsatisfactory.

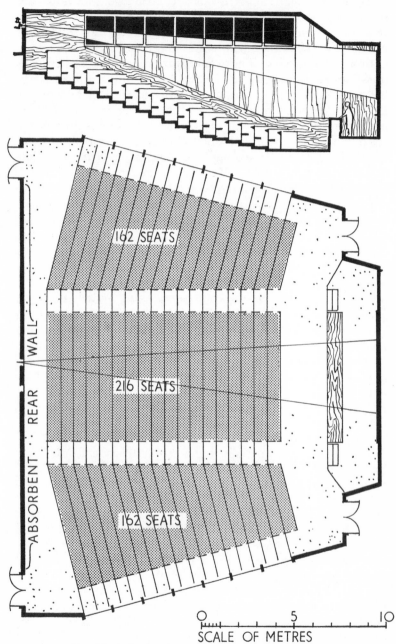

FIG. 27. A Design for a Lecture Theatre to seat 540

Attention is particularly drawn to the value of a reflecting splay protruding into the auditorium from the top of the proscenium opening, and indeed to the correct employment of useful reflecting surfaces in the whole ceiling design.

Emphasis is laid on the value of carpets in theatres. Not only do they provide useful sound absorption but also reduce intrusive audience noise. Their omission is often a false economy. It is sometimes necessary to locate some absorbents on the side walls near the stage end of the auditorium. This reduces cross reflections and reverberation in this area, both of which can adversely affect intelligibility in the front rows of seats.

LAW COURTS

Difficulties in hearing in Law Courts are frequently reported. This is probably largely because witnesses and prisoners are naturally reluctant to speak up. The importance of ensuring that the court is well insulated from extraneous noise cannot be overemphasised, particularly when the site is surrounded by busy streets. The court should be planned so as to bring the witness box and bar as close as possible to the magistrates' or judge's bench and the jury seating. Very high ceilings should be avoided, and the volume of the room and reverberation time should be adjusted to the values recommended earlier. Resilient floor coverings in public spaces and general circulation areas will ensure a minimum of foot traffic noise.

CLASSROOMS AND LECTURE THEATRES

The acoustical design of small classrooms, say up to about 100 m² in floor area, depends very much on the nature of the building structure. If it is one of the recent types of building with light-weight panel walls, very large window areas and a light suspended ceiling, then there will very likely be no need to incorporate any special sound-absorbent treatment, or at most a small amount of acoustic tile or some similar absorbent material at the sides of the ceiling. A rough check of the average reverberation time can be made and if this is in the region of 1·0 to 1·5 seconds (with the room empty) the conditions will be suitable. On the other hand, if the building is of heavier and more soundreflective materials, e.g. brick walls, solid concrete ceiling, etc., then in order to reduce excessive reverberation (which can rise in this type of building to a value of up to 4 seconds) some sound absorbents must be incorporated. The ceiling is the most tempting surface for the application of sound absorbents, largely because most of the economical treatments are not suitable for use on walls where accidental or malicious damage to them can occur. However, the middle part of the ceiling is a useful reflector, and the absorbent treatment should, therefore, be kept to the edges, and if necessary the upper part of the walls.

Lecture theatres for audiences of up to at least 500 should not require amplified speech if they are acoustically designed on the

TABLE II

REVERBERATION TIME CALCULATION FOR A LECTURE THEATRE (SEE FIG. 27) BY THE SABINE FORMULA

Size: 1530 m³.

Seating: 540 (2·83 m³ per seat).

Recommended reverberation time (from Fig. 26): 0·8 sec approx. This value required for two-thirds full audience.

Surface	Area, m^2	Hz, Coefficients and Sabins					
		125		500		2000	
		α		α		α	
Floor, cork tiles on concrete (including risers)	410	0·02	8	0·05	20	0·1	41
Ceiling, suspended lath and plaster	330	0·3	99	0·1	33	0·04	13
Window, 6 mm plate glass	29·8	0·1	3	0·04	1	0·02	1
Ply panels with rock wool in cavities on walls	201	0·4	80	0·15	30	0·1	20
Perforated (10%) plywood 25 mm rock wool (rear wall)	46·5	0·4	19	0·9	42	0·75	35
Seats, padded	540 (No.)	0·08	43	0·15	81	0·18	97
Air	1530 m³	—	—	—	—	0·007	10
Totals		252		207		217	
Two-thirds audience	360 (No.)	29		90		94	
Total		281		297		311	
Reverberation time (sec)	Empty	1·0		1·2		1·1	
	Two-thirds audience	0·9		0·8		0·8	
	Full	0·8		0·7		0·7	

correct principles given above. A typical example of a suitable design is given in Fig. 27, and this design is analysed and all calculations of the reverberation time at 125, 500 and 2000 Hz are given opposite.

INFLUENCE OF AMPLIFICATION SYSTEMS

When a room is to be served entirely or predominantly by reproduced sound from a loudspeaker system, whether it is amplified speech from a microphone or reproduction of recorded material as in a cinema, there is less reason to observe the requirements of natural acoustics and the design is simplified. The amplifying system will invariably be able to supply an abundance of acoustic power. There is therefore no call for supplementing the direct sound by reverberation for speech, and we have seen from Fig. 23 that intelligibility tends to increase as the room is made more dead. Recorded music will generally have an appropriate amount of reverberation embodied in the recording (from the studio in which it is made) and there is thus little to be gained by adding further reverberation to this. The design aim for the room should be to provide a reverberation time rather less than the speech optimum. The acoustic 'deadening' of the room must not be taken too far or a claustrophobic effect will be produced.

The size of the room should be kept in the range 3·5 to 4·2 m³ per seat, preferably nearer to the lower value. The shape of the room should be such as to avoid any danger of long return paths for sound, giving echoes which can arise just as easily in an acoustically 'dead' room as in one with longer reverberation.

The position for the loudspeaker in a cinema is automatically settled. It must go behind, or very close to, the screen. In other rooms for reproduced sound a similar position should be adopted. When the loudspeaker is to be fed with amplified speech from a microphone in the room itself, then the problem of feed-back arises, and the whole matter becomes more involved, but architects are advised to consult a speech-reinforcement engineer at an early stage so that the loudspeakers can be integrated into the room design rather than added later as appendages.

4

The Design of Rooms
for Music

The acoustical environments in which music has been performed have had a large influence on the whole art, both in the historical development of music and in its present appreciation. Man's interest and enjoyment in the contributory effects of reverberation have probably derived from his centuries long background of living in buildings and may indeed go back to the prehistoric days when he first took to occupying caves. This would suggest that those peoples who have persisted in an open-air or nomadic life might be less likely to appreciate the acoustical effects of rooms, and reflection on the types of music made by such peoples seems to confirm this.

Certainly, the recitation of the liturgy in very large basilican churches was influenced by the acoustical conditions; the long reverberation time made some form of musical intonation irresistible. Similarly, the development of the Italian opera-house with its elaborate ornamentation, plush furnishings and tiers of boxes—resulting in a short reverberation time—provided the suitable acoustical conditions for the rapid music written by Mozart and his immediate predecessors. (Another influence on the development of opera-houses was the tradition for attendance to be a social occasion, see Fig. 28.)

Until about the beginning of the nineteenth century music had its appropriate auditoria—church, opera-house, salon—and was written in styles which, consciously or unconsciously, had been moulded by the acoustics of those auditoria. However, at about that time musical and social changes led to the present state of affairs where music is no longer conceived in terms of the acoustical environment of any particular building but instead now demands that buildings be created to suit its own requirements, a step first taken in practice in 1876 by Wagner at Bayreuth.

FIG. 28. Royal Opera House, Covent Garden, London

Nevertheless, the acoustical design of rooms for music is still strongly based on tradition. This is not to say that the tradition cannot change. In music, as in all things, a developing process is always present. Nor is tradition necessarily of one kind, because as we have shown music is of many kinds and has been heard in a wide variety of circumstances. Added to this is the diversity of opinion found in all artistic matters, each opinion no doubt being influenced by the personal experiences of its holder.

The present state of knowledge about the acoustics of rooms for music is such that major faults (such as echoes) can be avoided in the design or, failing this, can be eliminated by suitable remedial measures in the completed building. It is thus possible to design a room which has good acoustics but it is not possible to be sure of designing a room with excellent acoustics. One reason for this is that there is no firm agreement among musicians as to what constitutes excellent acoustics, and in any case acoustics which are excellent for, say, romantic music will not be excellent for classical music. Another reason is that acoustical design is a matter of compromise and the degree of compromise that is necessary will depend on the circumstances. For example, it may be necessary to sacrifice some of the acoustical qualities in order to ensure reasonable conditions for the audience at the back of the room.

Further, as will be obvious in this chapter, nearly all the advice that can be given is qualitative only, at this stage of knowledge. The one important exception is the reverberation time which can be specified (at least within a range of values) and which can be calculated beforehand with a fair degree of accuracy. Again, the reverberation time is the only acoustical quality that can be measured objectively (if we exclude measurements of noise levels). But if obvious faults are found to exist in the completed building then their causes can be tracked down instrumentally.

We will now discuss the musical qualities that are desirable, and then describe the design factors that affect them. Lastly, the different classes of rooms for music will be considered.

Musical Requirements

The musical requirements (not in order of importance) that are affected by the acoustical design of rooms are: (i) definition, (ii) fullness of tone, (iii) balance, (iv) blend, (v) no obvious faults, such as echoes, and (vi) a low level of intruding noise. Further, it is desirable to obtain reasonably uniform acoustics over the whole audience area.

Some description of these musical terms is necessary. 'Fullness of tone' is the most difficult to define although it is easily recognised. Perhaps all we can usefully say about it is that it is the satisfying quality added to the sounds produced by musical instruments (or voices) when in a room as compared with in the open air. Although there may be subtle differences we must assume for design purposes that musicians mean nearly the same quality when they use such terms as warmth, richness, body, singing tone, sonority or resonance. 'Definition' has two main characteristics; the first is concerned with hearing clearly the full timbre of each type of instrument so that they are readily distinguished one from another; the second is concerned with hearing every note distinctly so that, for example, it is possible to hear all the separate notes in a very rapid passage. (Speeds of playing of 15 notes per second are not uncommon.) This implies that the sounds from the whole orchestra should be heard well synchronised. 'Clarity' is a term commonly used as an alternative to 'definition'. 'Balance' we would define as the correct loudness ratio between the various sections of an orchestra as heard by the audience. 'Blend' is another quality difficult to define but in general terms it is the possibility of hearing a body of players as a homogeneous source rather than as a collection of individual sources.

All these qualities are important in all types of rooms for music but the emphasis is different. For example, fullness of tone may be more important than definition in a concert-hall, while in an opera-house the reverse may be true. This is discussed below under types of auditoria.

Design Factors

For design purposes the most convenient way of regarding the behaviour of musical sounds in an auditorium is as follows. The sound from any given instrument or voice will spread out more or less uniformly in all directions. (Actually, most musical sources are directional at some part of the frequency range. However, this need not concern us here except for one example which is discussed below.) As the distance from the source increases the loudness of the direct sound (i.e. the sound travelling directly from the source to the listener) decreases due to the spreading of the sound (the inverse square law), and roughly speaking the loudness will be halved for every trebling of the distance. Thus for example the loudness of the direct sound for a listener 30 m from the orchestra will be half that for a listener about 10 m from the orchestra. Further, the loudness of this direct sound will be still more reduced if it has to travel at grazing incidence over the heads of the audience: the sound waves tend to be 'sucked in' towards this absorbent surface.

All reflections of the original sound which arrive at a listener very shortly after the direct sound will to him be indistinguishable from the direct sound: the sounds will coalesce into a whole and the total loudness will be increased. How soon after the direct sound a reflection must be to coalesce is not known accurately but probably it should not be more than 35 milliseconds later. The speed of sound is about 340 m per second so this 35 milliseconds corresponds to a path difference of about 10 m. For convenience we will call all the sound reflections that arrive at the listener within this 35 milliseconds the 'first-reflected' sounds, although some of them may have been reflected more than once.

After this initial period there will be many reflections of the sounds arriving at the listener and all getting steadily less loud because they have had to travel longer and longer distances round the auditorium to get to the listener and will have been reflected off partly absorbing surfaces on the way. This is the general reverberation process. Any particular reflection which is louder than its immediate neighbours in time will tend to be heard as an echo: the more it 'sticks out', so to speak, above the general reverberant sound the worse it will be.

The process is illustrated in Fig. 29, which shows the direct sound followed in the 35-millisecond period by five 'first-reflected' sounds all of which will blend with the direct sound. After these comes the general reverberation with, in this example, two echoes. Note that the second echo will probably be as bad as the first because although its absolute loudness is less it 'sticks out' above the reverberant sound level by as much as the first one does.

(i) Definition

The two requirements, 'definition' and 'fullness', are inter-related. As for speech, definition depends on the listeners receiving the direct sound and first reflections of it arriving not more than 35 milliseconds later at a strength well above the reverberant sound level. On the other hand, fullness appears to depend mainly on having plenty of reverberant sound. Thus it is likely that definition will suffer if the reverberation is made sufficient for maximum fullness, although there may be a range over which fullness can be varied without any appreciable loss of definition.

Another point is that the loudness of music depends largely on the direct and first reflected sounds. Thus if sufficient loudness is to be maintained over the whole audience area then the direct and first reflected sounds must be maintained at as high an intensity as possible.

The audience near the front of the hall will obviously receive the direct sound loudly, but as we go towards the back of the hall the intensity of this direct sound will fall off. Nothing can be done about the loss due to the inverse square law, but the

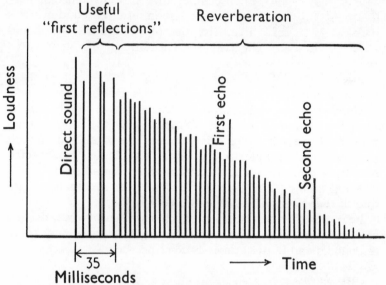

FIG. 29. Illustration of Behaviour of Sound

grazing incidence loss can be kept to a minimum by raking the seats. How to design the rake is described in Fig. 24, and for music the 'free height' between successive rows of seats should be 100 mm if possible. Further, the longer the hall the more the back rows are going to lose definition, and it is obviously desirable to keep the length to a minimum. This is one advantage that fan-shaped or horseshoe-shaped halls have over rectangular halls.

In the larger halls, say those seating about 1500 and upwards, the definition and the loudness at the back will not be very good unless the first reflected sounds are used to reinforce the direct sound. For example, a reflector over 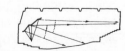 the orchestra can be shaped so that it reflects the sounds towards

85

the back of the hall. The sounds travelling by this path will still be reduced by the inverse square law but will no longer be at grazing incidence. Thus their intensity when they arrive at the back of the hall may be the same as or even greater than the intensity of the direct sound and they will help to maintain definition and loudness at the back. Similarly, the ceiling may be so shaped as to direct sound towards the back.

If there is a balcony the listeners under it will be shielded from  many of the useful reflections from the main ceiling, and thus any reflector over the orchestra should reflect sound into this area. Also, the soffit of the balcony should be used as a reflector.

(ii) *Fullness of Tone*

This quality is governed mainly by the reverberation time—the longer the reverberation time (within reason) the better the chance of obtaining adequate fullness. What the reverberation time should be is discussed below under types of hall, but fullness is also affected by the strength of the direct and first reflected sounds. It appears that if the direct sound is assisted by strong first reflections then the reverberation time should be longer than if there were little first reflected sound.

It is possible that fullness is increased by cross-reflection between parallel surfaces. However, such surfaces may cause resonances and flutter echoes as described below.

(iii) *Balance*

This quality is partly under the control of the conductor but the platform or orchestra pit design will also affect it. It is obvious that the weaker instruments such as the woodwind should not be still further weakened by, for example, being hidden behind the strings. On the other hand the strongest instruments such as the brass will come to no harm if so screened. These points are discussed further under platform design (for concert-halls) and under orchestra pit design (for opera-houses).

(iv) *Blend*

Again, this quality is partly under the control of the conductor but to a lesser extent than is balance. The main factor in room design that can help blend is the provision of reflecting surfaces close to the orchestra. Under these conditions the sounds are to a certain extent 'mixed up' before they reach the audience. Added advantages are that the players can hear their fellows clearly and that support is given to their own playing. Further, if the orchestra platform is too wide the sounds will arrive at the listeners from widely divergent directions and this detracts from the blend. This also is discussed further below under platform and pit design.

(v) *Faults*

There are three acoustical faults that can ruin an otherwise good room and they are: echoes, resonances and flutter echoes.

The terms 'echo' and 'resonance' are sometimes used by musicians as synonyms for reverberation, but here by 'echo' we mean a repeat of the original sound coming so loudly and so long afterwards that it is heard as a separate entity, and by 'resonance' we mean the accentuation of a small frequency band of sounds. This latter will cause these frequencies both to be louder and to die away more slowly than the rest of the frequency range.

Thinking in terms of geometric acoustics an echo is caused by a reflection from some surface. The seriousness of an echo is determined by how long it is after the original sound and by how loud it is compared with the original sound—the further behind it is and the louder it is the worse it will be. Two other points are, first, that the longer the reverberation time the less serious any particular echo is likely to be, because it will tend to be 'covered up' by the reverberation; secondly, that the higher-frequency echoes are likely to be more serious, mainly because the ear is more sensitive to echoes at the higher frequencies but partly because the shorter the wavelength of the sound the closer it will follow geometric laws of reflection.

Consider for example the reflection off the rear wall. The audience near the front of the hall will receive the original sound loudly, and the reflection from the rear wall will be a good deal less intense than this because it has had to travel to the rear of the hall and back to the front again. On the other hand, for these listeners the time interval between the arrival of the original and reflected sounds will be consider-able because of the long path difference (i.e. the difference between the path taken by the original sound, which is just from the orchestra to the front part of the hall, and the path taken by the reflected sound, which is  from the orchestra to the rear of the hall and then back to the front). The echo will be worse, in this example, if the rear wall is concave—as it would usually be in a fan-shaped hall—because this would increase the loudness of the reflected sound at the front of the hall by focusing. By contrast, for listeners at the back of the hall the reflection off the rear wall will be comparable in loudness with the original sound but the time interval between the arrival of the two sounds will be less. If the rear wall was made absorbent then the intensity of the reflected sound would be lessened and the risk of echo from this surface reduced. (If the rear wall were perfectly absorbent at all frequencies—which is not practicable—then there would be no risk.) Similarly, if the exposed area of the rear wall, i.e. the area above the heads of the audience, is small then the intensity of the reflected sound will be less and the risk small. An alternative treatment for the rear wall would be to make it diffusing.

The rear wall has been discussed as a simple example but obviously other surfaces of a hall may also cause echoes. Further, a twice-reflected sound (e.g. via rear wall and ceiling) might also be loud enough to cause an echo. As discussed in Chapter 2, any reflecting concave surface is dangerous but the greater the radius of curvature the less the focusing and thus the less the danger. Further, if a concave surface focuses only in space where there is no audience the danger is obviously lessened. It should always be remembered that the design of a room is a three-dimensional problem. The design methods must rely on geometric acoustics, but this is a simplification of the actual behaviour of sound and it may be that echoes will occur in completed halls which cannot be explained in terms of geometry. In such cases the only course is to investigate them by instrumental methods, e.g. directional microphones, but this is beyond the scope of this book.

The second type of fault, resonances, occur when some small frequency band is 'favoured' by the room shape. An example is two parallel and reflecting surfaces; a sound whose wavelength is exactly equal to the distance between the surfaces, or to a sub-multiple of it, will tend to be louder than sounds of other frequencies. (This phenomenon is known as 'colouration' in studio acoustics, and can be likened to the behaviour of a violin string which when bowed will vibrate at a frequency determined partly by its length.)

The last type of fault—flutter echoes—is again likely to occur between parallel reflecting surfaces but the mechanism is different. In this case any short burst of sound will travel to and fro between the parallel surfaces and if the surfaces are far enough apart will be heard as a series of echoes diminishing in intensity. However, while flutter echoes will be generated between any reflecting parallel surfaces which are at least 15 m apart if the sound is made in the area between the surfaces, when the source of sound is at some other place then the flutter echo will either be very much reduced or will disappear. For example, in a concert-hall with parallel walls a strong flutter echo may be heard if the hands are clapped in the audience area between the walls, but the orchestra, not being between these walls, is not likely to excite a serious flutter echo.

Both resonances and flutter echoes are only minor risks, and neither will occur if the surfaces are out of parallel by as little as 5°, or if one of them is absorbent. As resonances are more likely to occur at low frequencies the surface must be a good low-frequency absorbent; flutter echoes are more likely to occur at mid and high frequencies, so in this case the absorbent must be effective at these frequencies.

DESIGN OF AUDITORIA

We now give advice which is as practical as possible on the design of the various types of auditoria used for music, namely concert-halls (including recital halls), opera-houses, churches and cathedrals, multi-purpose halls, rehearsal rooms and music rooms. However, as we have already said and as will become more obvious, the advice must be so general and so much compromise is necessary that a good deal of judgement will be called for in any particular design. This judgement can only be based on experience and this indicates that a consultant should be employed whenever possible.

CONCERT-HALLS

(i) Shape

The three basic shapes are rectangular, fan and horseshoe. Other shapes are possible; the Royal Albert Hall is oval and a circular concert-hall has been proposed but they are not common and have no tradition behind them. They may thus have faults which are not obvious on paper and which would be exposed only in practice.

Dealing first with the horseshoe shape, although this is the traditional shape for opera-houses its suitability for concert-halls is dubious. The main reason is that the walls must be sound absorbent, either by using absorbent materials or by covering them with tiers of galleries or boxes, and this will result in a reverberation time too short for orchestral performances (but not for opera). Nevertheless, this shape has been successfully used in the past, e.g. the Usher Hall, Edinburgh, but then the seating was extremely plain, thus tending to keep the reverberation time up. With the modern demand for upholstered seats it is doubtful if a long enough reverberation time can be got with this shape.

The main disadvantage of the fan-shaped hall is that the rear wall, balcony front and seat risers are all curved, causing a serious risk of echoes. The rectangular hall is freer from this risk and in addition has a possible advantage that there is more cross-reflection between the parallel walls which may give added fullness. (But with parallel walls there are the minor risks of resonances and flutter echoes.) The weight of tradition is on the side of rectangular halls, but then it is only in very recent times that very large audiences have had to be accommodated and the main advantage the fan shape has over the rectangular is that the length can be less. Thus the difference between the intensity of the direct sound at the front and at the back will be lessened, giving greater uniformity of acoustical conditions.

Modifications to the basic rectangular and fan shapes are of course possible. One way to reduce a main disadvantage of the

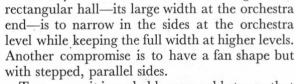

rectangular hall—its large width at the orchestra end—is to narrow in the sides at the orchestra level while keeping the full width at higher levels. Another compromise is to have a fan shape but with stepped, parallel sides.

To sum up, it is probably reasonable to say that if fullness of tone is the prime requirement and that risks can be taken to achieve it then have a rectangular shape; if definition is the prime requirement and fullness is secondary then have a fan shape and make all the curved surfaces very absorbent.

Whatever the plan of the hall the audience seating should be raked as shown in Fig. 24. It is better to keep accurately to the designed rake than to approximate to it by one or two straight slopes. The reason is that these approximations will tend to reduce the free height for those at the back who most need strong direct sound. However, the step rise will vary and this is a nuisance. Further, a limit will be set to the maximum rake by the safety regulations.

The audience rake will depend on the platform height. Platform design is discussed below, but we will say here that the front part of the platform should not be less than 0·6 m nor more than 1·2 m. If it is less than 0·6 m then the 'command' which the musician—particularly the solo singer—likes to have over the audience tends to be lost; if it is higher than 1·2 m the centre section of the orchestra will be screened—both acoustically and visually—from the front rows of the audience by the front rows of the orchestra.

In a large hall some form of balcony will be almost essential—otherwise the length of the hall will be excessive. The rake of the balcony should also be properly designed. In general the audience in the balcony will get plenty of direct sound because the sound is not at grazing incidence over the stalls audience. Further, the orchestral reflector (if any) and the ceiling can easily be designed to reflect sound to this area, but under the balcony the conditions are more difficult. This is because the direct sound is at grazing incidence, because this area is cut off from a large part of the ceiling and because a reflector cannot be designed (see below) to direct sound from all parts of the orchestra into this area. For these reasons the depth of recess under the balcony should be kept to a minimum and should never be more than twice the free height (i.e. the height between the heads of the audience and the balcony soffit) at the entrance to the recess. Also, the balcony soffit should be reflecting and preferably shaped as illustrated on p. 86. There is a slight risk of echo from the rear wall under the balcony, particularly in a fan-shaped hall, so for safety it should be made absorbent. However, this will make the conditions rather 'dead' for the back two or three rows of the audience and a better solution is to cant forwards the top section of the wall; this not only gives a useful reflection for the back rows but will also prevent an echo.

This discussion of the balcony soffit profile should lead us to the consideration of the main ceiling area and the area over the orchestra, but it will be more convenient to discuss first the design of the platform (mainly after H. Creighton). We shall see that the first objective in design should be to keep the area of the platform as small as possible, so that it is better to plan for the normal maximum numbers and to leave exceptional demands (e.g. for massed choirs) to be met by temporary expedients. The normal maximum for a large hall at the present time may be taken as: orchestra 95–100, plus two pianos and four vocal soloists; choir 300–350. For all except the most important halls these figures could be reduced to 85 and 200 respectively.

The important dimensions for planning are:

1. A seated player of a violin and of most wind instruments needs an area 1 m × 0.6 m—horns and bassoons rather more.

2. A tier 1.1 m deep is sufficient for all string and wind players, including 'cellos and double basses; players, however, prefer at least 1.2 m.

3. Tympani and percussion need a tier 2 m deep.

4. Risers to tiers should not exceed 0.5 m because of the difficulty of carrying heavy instruments up them.

5. A piano (concert grand) measures 2.75 × 1.6 m on plan.

On the basis of these figures a reasonable allowance of space for the platform of a large hall would be: orchestra, 93 m² choir, 140 m²; total, 233 m².

The dimensions of the platform could therefore easily be as great as 15–18 m deep or 24–28 m wide, and the sound paths from the more remote performers to some listeners could exceed those from the nearer by these distances. Reflections from walls surrounding the platform might be delayed by twice as much. The result would be apparently ragged ensemble and poor definition even when all the performers were playing strictly on the conductor's beat. Not only this, but the delays would affect performers also, so that their more distant colleagues would seem to be off the beat, and sensitive combination and accurate intonation would be more difficult to achieve.

There is reason to believe that, for ideal musical conditions, sound paths from different instruments should not differ by more than about 10 m. Clearly a concert platform must exceed 10 m in one or both dimensions, but the fact that ideal conditions are not in any case practicable emphasises the desirability of the most compact arrangement possible.

Platform dimensions affect tone in another way also. Blend depends in part upon the apparent size of the orchestra to a listener—that is to say, the angle within which his view of it is contained: if this angle is too wide, sound comes to him from widely divergent directions and blend is destroyed. This would seem to suggest that a narrow platform is better than a wide one; but if it is narrow it is also deep, so that path differences are at a maximum for all parts of the auditorium.

These points are illustrated diagrammatically for various

possible shapes of platform and for listeners in centre and extreme side seats at the front and 12 m back. At this distance the inevitable defects of front seats should have disappeared. Path differences from the whole platform are figured at various seating positions, but it is from the orchestra alone that they are most important; choirs cannot in any case achieve the precision and speed of instruments, so that they can tolerate longer delays. Taking a maximum path difference of 11 m from the orchestra alone as a reasonable aim, the shading shows those parts of the auditorium where this figure is exceeded. For blend an angle of 60° is a useful, though very imprecise, indication of the nearest seat from which it is likely to be regarded as satisfactory.

On the basis of these tentative criteria the shallow, wide platform (A) gives really bad path differences in front side seats; they improve rapidly towards the rear and centre, where they are the shortest possible. Blend is, however, still bad 12 m back. The deep platform (B) gives good blend, but path differences affect all seats and are excessive both for the whole platform and for orchestra alone. (C) is a compromise; the variation from side to centre is evened out somewhat as compared with (A), while the uniformly excessive path differences of (B) are avoided; the conditions assumed to be satisfactory for blend are

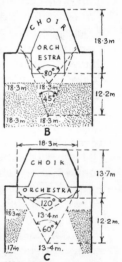

achieved at 12 m from the platform. Something like this represents the best arrangement for a platform of the size we are considering; it is planned within a rectangle 18 × 14 m. The depth of 14 m should be taken as an absolute maximum, and if the width of 18 m is exceeded it should be only by those choir seats which are least frequently occupied. But for smaller platforms it is more advantageous to tone to reduce the width than the depth.

In longitudinal section an orchestra platform can be flat, flat for the front part and stepped for the back part, or completely stepped from front to back. With the flat design some of the weaker instruments—the woodwind, the violas, the 'cellos—are screened by the players in the front, and so are the strongest instruments—the brass and percussion—but these can stand screening. The demands of pianos have probably perpetuated

what is the most frequent, but acoustically the most unsatisfactory, kind of platform—that which has a large flat in front
and a few tiers at the back. In this case only those instruments
which do not need it—brass and percussion—have the advantage of exposure, while the majority of strings and woodwind
screen one another's sound on the flat.

It is probable that the best design is to have a fully-stepped
platform. It is true that the powerful instruments are still un-
necessarily exposed, but with all the instruments given an equal
chance it can then be left to the conductor to achieve the correct
balance.

The platform will be made of wood on a wooden frame; this is
traditional and essential for the 'cellos and double-basses, some
of whose sound is due to radiation from the platform. But it may
be better to make the back tier holding the percussion of wood
on solid concrete, because this will reduce their sounds a little
which otherwise tend to be overpowering. However, some works
are probably intended to be overpowering, so perhaps the percussion should be given full scope: for other works the balance
can be left to the conductor.

Some platform space can be saved if holes for the music stands
are provided in the nibs of the risers, instead of the usual tripods.

There will usually be a barrier between the orchestra and
choir because when there is no choir these seats will usually be
occupied by audience. (Acoustically, this is a little undesirable
because they will provide absorption where none is wanted.)
The barrier should be removable so as to provide greater flexi-
bility of lay-out when there is a choir. The risers for the choir
tiers need to be no more than 180–200 mm.

Conductors often prefer a rostrum that is off the platform
altogether, and it ought to be capable of adjustment to suit
individual preferences.

The placing of a pipe organ is always a problem because of
its size. (Electronic organs present no problems but are often not
acceptable to musicians.) The pipe organ can either go behind
the orchestra, where it will then sound as a homogeneous whole,
but with a raked platform it will tend to make the back of the
ceiling too high above the main floor level and complicate the
design of the ceiling area over the orchestra. If placed at one
side or in two halves on either side then it will not sound as a
whole, although this defect can be minimised by keeping all the
pipes of a particular stop wholly to one side or the other. In
either position it will provide unwanted absorption for the

orchestral sounds. On balance it can be said that if some considerable sacrifice of the orchestral sound is acceptable then the organ should go at the back; if the minimum sacrifice is required it should go at the sides. In either case the console should be movable to suit either concertos or solo recitals.

We will now consider the surfaces round the orchestra: the rear wall, the sides and the ceiling area over. In principle these surfaces can fulfil one of three purposes: to absorb the sound, to reflect it in random directions or to reflect it in specified directions. It is generally agreed that to absorb the sound is harmful; not only do the players not hear themselves but a lot of useful sound energy is being wasted which could otherwise be going to assist the direct sound in the auditorium. This is a particularly serious loss in large halls: orchestral instruments have not been developed so as to produce more sound, nor, usually, are they too numerous, while audiences continue to increase in size. It seems a pity, therefore, to waste such energy as is available. Nevertheless, it is often the case that the sound is absorbed, usually by curtains draped round the sides of the orchestra.

If, then, these surfaces are to be reflecting, should they be used to direct the sound in specified directions or should they provide random reflections? There is no doubt that in the larger halls— say those seating 1500 to 2000 and upwards—the definition and loudness towards the back of the hall will be much less than towards the front unless these reflections are directed towards them, even although the ceiling is shaped so as to help them. On the other hand, those directed reflections—some of which are bound to reach all of the audience—may detract from the fullness of tone unless the reverberation is made particularly long, which may not be possible. If these surfaces are made random so that the sound is diffused as much as possible then this disadvantage is lessened, but at the expense of the audience at the back. For effective diffusion, projections of the order of size of at least 0·3 to 0·6 m are necessary (see, for example, Plate I); an alternative is large-scale convex surfaces, in long section and possibly also on plan. In either case, if the surfaces are formed on frames, thus enclosing air-spaces, they must be of a good weight and thickness—say a minimum of 12 mm solid wood or plaster— or else they will provide low-frequency absorption.

Diffusion round the platform also ensures that some sound is returned to the players, so that they can hear their own instruments and those of their colleagues better, and if directional reflectors are necessary parts of them should be flat for the same purpose.

The area over the platform (see, for example, Plate II) either can form part of the main ceiling line or a suspended orchestral reflector can be used (as in the Royal Festival Hall, see Plate III). The acoustical advantage of the suspended reflector is that the space between it and the actual ceiling forms a useful addition to the total volume of the hall (which, as we shall show below, must reach a certain minimum value for adequate reverberation).

Whether the ceiling is shaped or a separate reflector is used the requirements in a large hall are, then, to reflect sounds towards the rear of the hall—these reflections to follow the direct sound by as short a time interval as possible—and to reflect some of the sound back to the orchestra to help the players hear themselves and each other. The second requirement can be met by making parts of the area horizontal, but the main requirement is complicated by the large area covered by the sound source. It is clear that an angle for the reflecting area suitable to reflect instruments at the front of the orchestra towards the rear of the hall is not suitable for instruments at the back, and vice versa.

Consideration of the design actually used for the suspended reflector in the Royal Festival Hall will help to illustrate the problems (see Fig. 30). The height of the rear leaf of the reflector in this case was set by the demand of the organ consultant for a free opening for the organ 10 m high. This leaf reflects sound from the choir and the rear instruments of the orchestra towards the rear of the hall. On the other hand, sound from the front instruments is reflected straight back to them. If the angle of inclination to the horizontal of this leaf were increased to reflect the front sound more towards the rear of the hall, not only would the rear instruments be reflected to the ceiling but the other leaves of the reflector surface would have to be higher since the back point is fixed in space. The front leaf reflects the front instruments' sound towards the rear but the back instruments' sound towards the ceiling; a reduction in the angle of inclination in order to help the back instruments would tip the front instruments' sound too much towards the floor area.

There is a further consideration: while at middle and high frequencies the dimensions of each leaf of the reflector are large compared to the wavelengths and can therefore be considered as separate reflectors, at low frequencies the whole reflector must be considered as one. Thus if the angles of inclination of the

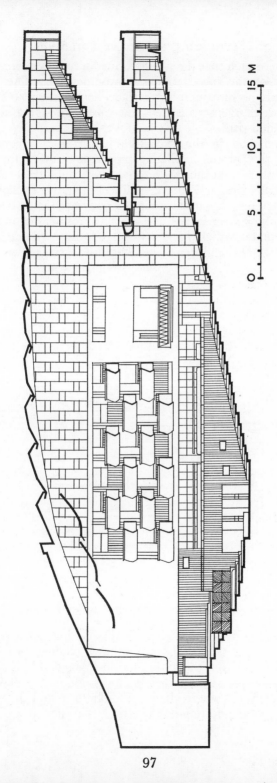

FIG. 30. Longitudinal Section of the Royal Festival Hall

97

leaves were such that the whole formed a concave surface, some focusing of the low frequencies would occur. The focusing might only occur at points in space where there is no audience, but we are not yet so confident in acoustical design as to be able to take such risks.

The reflector in the Royal Festival Hall is a compromise between the various conflicting requirements. The front leaf is at a slightly greater inclination to the horizontal than the middle leaf, which in turn is more inclined than the back leaf: the general contour is therefore slightly convex. It is made of wood 50 mm thick: the leaves are fixed by resilient mountings to timber beams which hang by tie rods from roof trusses. Its weight is 12000 kg plus 3000 kg of lighting fittings. This reflector

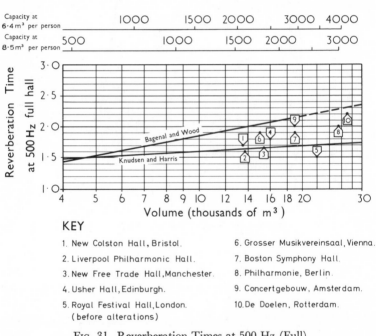

KEY

1. New Colston Hall, Bristol.
2. Liverpool Philharmonic Hall.
3. New Free Trade Hall, Manchester.
4. Usher Hall, Edinburgh.
5. Royal Festival Hall, London.
   (before alterations)

6. Grosser Musikvereinsaal, Vienna.
7. Boston Symphony Hall.
8. Philharmonie, Berlin.
9. Concertgebouw, Amsterdam.
10. De Doelen, Rotterdam.

FIG. 31. Reverberation Times at 500 Hz (Full)

was made as thick as this to ensure that the low frequencies were effectively reflected. Perhaps a thinner reflector would do in many instances, but the minimum thickness should be 12 mm.

The front row of the audience should not, for the players' and singers' comfort, be too close to the platform and the space left can serve as an additional reflecting surface, for the front row of players. It can be of any hard and polished material, e.g. wood, preferably on solid backing such as concrete.

(ii) *Reverberation Time*

The sole acoustical factor that is calculable at the present state of knowledge is the reverberation time. How to calculate it is described on p. 50 and how to measure it on p. 259; here we shall consider how long it should be and how to design a hall to get it, confining ourselves for the moment to the value at 500 Hz only.

Fig. 31 shows the measured reverberation times (at 500 Hz) of ten large concert halls, all with an audience present and all with reputations for good (if not necessarily excellent) acoustics. Also shown are the values recommended by Bagenal and Wood (U.K.) and by Knudsen and Harris (U.S.A.). However, it is clear from what we have already said that the choice of the best reverberation time is not a simple matter. First, if fullness of tone is the prime requirement then the reverberation time should be longer than if definition is considered more important. Secondly, if the hall is designed to provide strong first-reflected sounds then the reverberation time should be longer than if the platform-end surfaces are diffusive. Thirdly, romantic and choral music needs a longer reverberation time than classical or some modern music. Accordingly, in Fig. 32 we give a band of recommended values, but the modern tendency is towards the longer values.

The reverberation time is governed by the volume of the hall, by the amount of absorption in it and, to a small extent, by the shape of the room. The audience and the seating provide an unavoidable quantity of absorbent. For example, in the Royal Festival Hall full the audience and seating account for about 55% of the absorption at 500 Hz. In halls seating up to about 1500 there should usually be no difficulty in obtaining any reverberation time over the range shown in Fig. 32. If the volume per seat is about 5·7 to 6·4 m³, then for the longer values of reverberation time it will usually be necessary to omit all extra absorbent (at this frequency of 500 Hz)—the unavoidable absorbent will be sufficient. In the larger halls, however, and assuming upholstered seats, it is probable that, while a cube per seat of 6·4 or so will suffice to give the shorter values of reverberation time, to get the longer values the cube may have to be as much as 8·5. This raises two problems. The first is that, to get this cube, the dimensions of the hall will be so great as to create a serious risk of echoes and of course any absorbent treatment on particular surfaces to reduce this risk will also shorten the reverberation time. One possible solution is to suspend reflecting surfaces (the so-called 'clouds') below the main ceiling over most of the audience area. These will stop one cause of echoes—the long path for sound up to the ceiling and down again—without reducing the cube per seat being an extension of the ceiling surface as shown in Plate IV. Another possibility is to make the ceiling as diffusing as possible, in both plan and section.

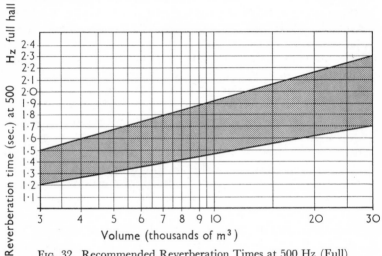

FIG. 32. Recommended Reverberation Times at 500 Hz (Full)

The second problem is one of cost. A hall seating 4000 at a cube per seat of 6·4 will only seat 3000 at a cube of 8·5. This is obvious, but as the profitability or otherwise of the hall will depend on the last 20% or so of the audience there is always a powerful force at work to reduce the cube per seat.

In modern halls there is often a suspended ceiling above which there may be a considerable volume housing some of the ventilation plant, lighting fittings, etc. The absorption coefficient of this ceiling will depend on its thickness and on the volume above it, and although this coefficient may not be much the area is so large that the total absorption may be considerable. In the smaller halls, or in the larger halls when only the shorter reverberation times are desired, the ceiling absorption will not matter, but in the larger halls and when the longer reverberation times are desired the ceiling absorbent must be kept to the minimum. Little is known about the coefficients of such ceilings but it is probable that the thicker and heavier the ceiling the less the absorption: 50 mm of solid plaster is desirable.

It should be pointed out that if the reverberation time is found to be too short in the completed hall and if the absorption has already been reduced to the minimum then nothing can be done to lengthen the reverberation. On the other hand, if the cube per seat is generous in the first place, not only may there be a margin in hand for acoustic treatment to stop echoes, if necessary, but if the reverberation time turns out to be too long (which is unlikely) it is a comparatively simple matter to shorten it. For example, if there is an area of wood panelling to provide low-frequency absorption (see below) some of this can be perforated to provide additional mid-frequency absorption.

It is often suggested that the reverberation time of a hall should be made adjustable to suit different kinds of music. However, we have already said that the major absorption will be due to the audience, and for the larger halls at least, although minor permanent alterations as instanced above are feasible, day-to-day alterations by significant amounts would involve adjustments to most of the wall and ceiling surfaces. Further, it is more probable that the reverberation time will never be long enough rather than that there will be some 'in hand'.

(Two possible systems of electronic reverberation are at the development stage. The first, 'ambiophony' uses microphones close to the orchestra; reverberation is added electrically, e.g. from an 'echo chamber' and then emitted from loudspeakers in the hall. The second, 'assisted resonance', selects a large number of the normal modes of air in the hall (see p. 47) and by electrical means lengthens the decay rate of each one.)

If there are any considerable volumes connected through openings to the volume of the auditorium (an example would be the roof space above a suspended ceiling with ventilation and lighting holes in it) then the reverberation times of these volumes should be made rather shorter than that of the auditorium. Otherwise the reverberation will be heard still continuing in these volumes after the main reverberation has died away.

For rehearsals and perhaps for broadcasts it is desirable for the reverberation to be not too different in the empty hall than when it is full. If upholstered seats are used the difference will not be too great, but can be reduced further by making the underneath of tip-up seats absorbent, e.g. by perforations with absorbent behind. However, this may cause a slight increase in the absorption of the seating even when occupied and might make it more difficult to get a long enough reverberation in the full hall.

Again, if it is required to keep the absorption of the occupied seats to a minimum then only those parts of the seat which are covered by the person seated should be upholstered, leaving the sides and back hard and reflecting.

The reverberant sound should die away smoothly, i.e. the sounds should decay at a constant rate and not in a series of

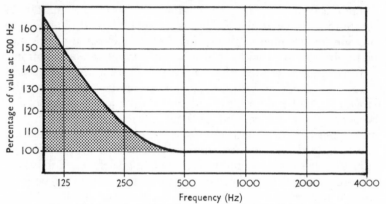

FIG. 33. Reverberation Time as a Function of Frequency

'jumps'. This requires effective diffusion of the sound throughout the room. While there will often be enough irregularities in the surfaces to ensure sufficient diffusion, when very large plane surfaces are used opposite each other some irregularities should be deliberately introduced on these surfaces.

We have so far been considering only the reverberation time at 500 Hz. At higher frequencies the air absorption begins to become important. Further, the absorption of the audience is considerable at the higher frequencies, and the problem is usually to maintain a long enough reverberation time at this end of the scale. At the lower frequencies the reverberation time can be the same as at 500 Hz, or for maximum fullness it may be better to increase it. Thus Fig. 33 shows the possible range of values at the lower frequencies compared with the selected value at 500 Hz.

The absorption coefficient of nearly all the surfaces in a hall and of the audience will be rather different at the lower frequencies. For the audience, for the seating, and for all 'soft' finishes on the solid the absorption coefficient at 125 Hz will be less than at 500 Hz; for all 'hard' impervious finishes on air-spaces the coefficient at 125 Hz will be more than at 500 Hz. While Fig. 33 shows that the reverberation time at 125 Hz may be up to 50% longer than at 500 Hz, if it is any longer than this it will be deleterious. On the other hand, if the reverberation time is the same at 125 Hz as it is at 500 Hz, or is even a little less, then although we may have lost some fullness the result will not be disastrous. In traditional constructions there was sufficient low-frequency absorption present in the form of fibrous plaster or other panels with air-spaces behind and of joist floors and of suspended ceilings to prevent this danger. But in modern constructions with solid walls, floors and ceilings it may be necessary to use particular low-frequency absorbents, e.g. wood panelling.

It is probable that a difference of 0·1 second in the reverberation time is the minimum perceptible when listening to music. Thus calculations should be made to the nearest 0·1 second, but the inaccuracies are such that the realised value may be as much as 0·2 second different—either way—and perhaps still more at the lower frequencies.

To summarise, the normal design procedure should be to decide on the required value at 500 Hz and to choose the volume of the hall and the amount of absorption in it to get this value. Preferably, and particularly in the larger halls, the volume should be chosen so as to get a rather longer time than the design value, so that in the completed hall the time can be shortened if necessary. Then the value at 125 Hz should be calculated and if found to be more than about 25% longer than the value at 500 Hz sufficient low-frequency absorption, e.g. wooden panels with an air-space behind, should be introduced into the design. Finally, the value at 2000 Hz should be checked. Although this will be determined mainly by the air and the audience it might be possible to change some surfaces, e.g. from curtains to perforated panels, to increase the reverberation (or vice-versa) at this frequency.

(iii) *Faults*

ECHOES

As we have said, the longer behind the direct sound an echo is, and the louder it is, the worse it will be. As a rough working rule path differences off large flat or convex reflecting surfaces should be kept below about 11 m, but in halls with the longer reverberation times path differences up to about 15 m will not be too serious.

The rear wall in any shaped hall is one of the most serious echo risks. Of course if there is only a very small area exposed— as will often be the case in a hall with raked seating and a balcony—then the risk is not so great. Further, in these cases it might be feasible to cant the wall forward slightly so that the sounds are reflected but towards the rear few rows only.

The second dangerous area is the junction of the ceiling with the walls. The right-angle thus formed can reflect sound back in the direction it came. This is particularly dangerous at the orchestra end, and in large halls a few feet at the ceiling margin should be absorbent, unless it is obscured by an orchestral reflector, or by 'clouds'.

The third area is the side walls at the orchestra end. This is one example where the directivity of instruments, in this case the brass, is important in the design. The brass emit more high-frequency sound to their fronts than they do to their sides. They will often be seated so that their sound is directed diagonally across the hall. Thus listeners near the front of the hall and seated on the same longitudinal line as the brass will receive the direct high-frequency sound rather weakly, because of the directivity, but the reflection from the side wall perhaps as strongly. The side walls can either be splayed so that the reflection goes further towards the back of the hall, thus reducing the path difference, or can be made diffusing or absorbent. Naturally, this risk does not arise in small halls where the width is such that the path difference will not be more than 9 m or so.

The last important area to consider for echoes is the ceiling. If the ceiling is shaped so as to reflect sound towards the back of the hall then the path difference is not likely to be enough to cause echoes. If, however, the ceiling, or part of it, is flat then the path difference may be too long. The worst danger is obviously near the front. Although such a flat ceiling can be made diffusing it should not usually be made absorbing because this will stop the useful reflections towards the back and will introduce too much absorbent. An alternative preventive method is to use the suspended reflecting surfaces ('clouds') previously mentioned and which shorten the path differences without reducing the volume.

RESONANCES AND FLUTTER ECHOES

The only point to add to the description of these faults given on p. 88 is that parallel surfaces, which may be desirable for fullness of tone, may cause these minor faults.

(iv) *Noise*

The noise criteria suggested for concert-halls are given in Chapter 10, and general advice on the reduction of external noise in Chapter 8.

The flooring under the seats should be chosen to minimise the noise of scraping feet. Carpet could be used but this will increase the absorption, probably undesirably. Cork or rubber are better because they will not increase the absorption.

(v) *Test concerts*

When a hall is completed one objective measurement—that of the reverberation time—is possible in the empty hall, and some listening tests, e.g. for echoes, can be done. However, a full test can only be made with an orchestra and with an audience present. These test concerts are desirable because, first, some faults, e.g. a resonance between parallel surfaces, may disappear when the audience absorption is present, or may get worse, e.g. an echo, when the reverberation time is reduced by the audience; secondly, if there are any such faults they should be sought out and corrected before the hall comes into regular use; and thirdly that some subjective assessment of the acoustics can be made and, perhaps, some adjustments made, e.g. reduction of the reverberation time, while the builders are still on the site.

The whole audience can be given questionnaires and asked for their opinions on such matters as fullness of tone, definition, echoes, etc., but it will be sufficient if groups of about 20 people each distributed throughout the auditorium are used. In addition it may be desirable to have professional musicians, e.g. music critics, in groups, and these groups should move to different positions in the hall at various stages in the concert so they can form an opinion of the conditions throughout the hall. It will alsc be very useful if a few specialist listeners move about the hall, i.e. people who know something both about music and about acoustics; they may be able to decide what it is that makes one position good or another bad. For example, they may be able to decide that an echo they can hear at one position is due to a particular surface.

OPERA-HOUSES

We have said that the traditional, or Italian, opera-house with its small cube per seat and large amounts of absorption resulting in a short reverberation time was eminently suitable for Mozartian operas. However, Wagnerian operas called for very different acoustical conditions—more fullness of tone, less definition, greater blending of the orchestra with the singers and some subduing of the larger orchestra with its augmented brass so as not to overpower the singers. These requirements led Wagner to design his own opera-house at Bayreuth. The main acoustical innovations were, first, the large volume per seat (nearly 8·5 m³) and the absence of tiers of galleries gave a reverberation time of about 2 seconds when full, i.e. almost twice the value of the average Italian type; secondly, the orchestra pit went some distance down under the stage and had only a restricted upward opening, thus blending all the instruments together and giving the correct balance between orchestra and singers.

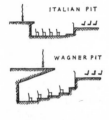

There are, then, two clearly defined types of opera-house. Which type is chosen will obviously depend on the circumstances. Such compromises between the two types as there have been do not appear to have been successful. However, it should be remembered that while Wagnerian operas can be performed reasonably well in the Italian opera-house, Mozartian operas are unsatisfactory when performed in the Bayreuth type. We shall therefore confine ourselves mainly to a description of the Italian type, with only passing references to the Wagner type.

(i) *Shape*

The traditional shape is horseshoe or semicircular. There is no acoustical reason why the rectangular or fan shape should not be used, but they have no advantage over the traditional shapes which bring the maximum numbers of audience as close to the stage as possible.

The seating should be raked as previously described, but as the stage platform will be flat or, better, only slightly inclined forwards the sight lines from the front rows to the back of the stage set one limit and from the back rows of the galleries or boxes to the front of the stage set the other limit. As for concert-halls the depth of the recess under a balcony should not be more than three times the free height at the entrance, and the balcony soffit should be shaped to reflect sound towards the back.

The traditional orchestra pit was only shallow and most of the orchestra was exposed but this may be one part of design where it may be better to depart from tradition. This is because orchestras have got larger and louder, not only for Wagner but also, for example, for Verdi. Thus it may be better to enclose—if not most of the orchestra as at Bayreuth—at least the louder sections of it. The interior surfaces of the pit should be reflecting but preferably of wooden panels with an air-space behind; if all the surfaces are solid there is a danger of too much low-frequency reverberation.

The size of the pit can be made adjustable to suit orchestras of different sizes. The average area required per player is about 1·2 m².

The ceiling of the auditorium should be flat or diffusing or, in the largest opera-houses, as far as possible shaped to reflect the voices towards the back. The domed ceiling has often been used in the past but always produces undesirable focusing and, if high enough, echoes. For example, in the Royal Opera House, Covent Garden (see Fig. 28), the domed ceiling focuses the voices from the back of the stage on to the top gallery, and the loudness of the voices varies considerably as the singers move about the stage.

It should be remembered that it is the singers, not the orchestra, who need most help and any reflecting surfaces should be designed with this in mind. A proscenium splay might be helpful, and certainly an apron stage, even if no more than 2 m from front to rear, is a useful reflecting surface. It should of course be boarded and not carpeted.

The horseshoe or semicircular opera-house will provide too much absorption for the Wagnerian conditions, and thus this type must be fan-shaped or rectangular. Fan-shaped is probably the better choice because it brings the audience closer to the stage. The design of the auditorium will then follow the same lines as discussed for concert-halls.

(ii) *Reverberation Time*

The reverberation time for the Italian type opera-house should be about 1·2 seconds (plus or minus 0·2 seconds) at 500 Hz when full. The volume per seat should be about 5·7 m³. For the Wagner type the reverberation time should be about 2 seconds, and this requires a volume of about 8·5 m³ per person. It is probable that, unlike concert-halls, the reverberation time at 125 Hz should be no longer than it is at 500 Hz. This is because the low-frequency reverberation will help the orchestra more than it helps the singers.

The volume of the stage is often as big as that of the auditorium. The amount of scenery and curtains in it will usually ensure that the reverberation time in the stage volume is no longer than that of the auditorium. It is most desirable that it should not be any longer, and if there is any doubt additional absorbent should be introduced into it. On the other hand, it should not be made too 'dead' because the singers will need some help from local reverberation. Unfortunately, because of the scenery requirements, it is not usually possible to provide what is acoustically desirable—namely, nearby reflecting surfaces for the singers.

(iii) *Faults*

The prevention of acoustical faults will be much the same for opera-houses as for concert-halls, except that flutter echoes and resonances are unlikely in horseshoe, semicircular or fan-shaped auditoria. All curved rear wall surfaces which are not hidden behind audience should be absorbent or diffusing, unless they focus in space where there is no audience. Echoes off the ceiling should be guarded against. The greatest danger is near the front, and a proscenium splay can be used or the ceiling can be made diffusing.

(iv) *Noise*

The same remarks apply as for concert-halls.

(v) *Test Performances*

The main advantage of test performances before an opera-house is opened would be to discover any faults. There is not so much need in opera-houses to attempt to balance definition against fullness as there is in a concert-hall.

CHURCHES AND CATHEDRALS

In *Planning for Good Acoustics* by Bagenal and Wood there is a full discussion of the acoustical requirements of churches and buildings for religious purposes and the reader is referred to this. The acoustical design will depend on whether speech or music is considered the more important element, and this will depend on the nature of the religious service. If speech is of first importance then the church should be designed in accordance with the principles given in Chapter 3. If a compromise between speech and music is required then the design should be as in Chapter 3 but the reverberation time should be lengthened to the lower values shown in Fig. 32 (p. 100). If the music is of first importance then the design should be as follows. (It should be mentioned that developments in speech-reinforcement systems have made it possible to make speech both intelligible and natural even under very reverberant conditions (see Chapter 6), although the necessary microphone technique imposes some limitations on the preachers.)

The long reverberation time (up to 7 or 8 seconds when full) of the medieval church led to the development of polyphonic choral music, and if this music is still to be performed then the reverberation time should be 4 seconds (at 500 Hz) when full. While this time will not be excessively long for later religious music (Bach and onwards) it would be better to have about 2·5 to 3 seconds. Also, cathedrals and large churches may be used for festival music of all kinds and the shorter reverberation time is more appropriate for this. The reverberation time at 125 Hz may be up to 50% longer than that at 500 Hz.

The volume per seat will usually be large enough to give this reverberation time or longer. Any additional absorption necessary should be distributed over the walls and ceiling areas and not concentrated in one place.

A church built with separate 'cells', e.g. chapels, opening off the main body of the church, has the acoustical disadvantage that the reverberation in the 'cells' may be different from that in the main body. Thus particular notes may undesirably continue in these cells after the main reverberation has ceased.

The choir will benefit from reflecting surfaces close to them so that they can hear themselves, and the higher they are placed the further will their sounds carry. Also, this will bring them closer to the ceiling which will return helpful sound to them. The common arrangement of the two halves of the choir facing each other with solid wooden screens behind them is also helpful.

Opinions vary on the best positions for the organ, but certainly it should not be far from the choir. Nor should it be narrowly confined in the organ loft. The console should be with the choir in the body of the church so that the organist can judge the total musical result. But there must be no obstruction between the organist and the organ, and from the console position he must be able to hear the full power and balance of the organ.

Extensive concave surfaces are best avoided as they cause a serious risk of echoes. If they are used then their radii of curvature should either be so large that they focus, so to speak, outside the building, or so small that they focus in space where there is no congregation. In general, the echo problem is not so critical in a church as in a concert-hall because the reverberation will be longer. Nevertheless, the west wall should preferably be absorbent. We have said that any extra absorbent required to control the reverberation should be distributed over all the surfaces, but this should not preclude its use for reducing echo risks from particular surfaces.

Finally we will mention one acoustical peculiarity and it is that a church with a long reverberation often has a 'sympathetic note'—that is to say, a region of pitch in which tone is apparently reinforced. This region lies between G and A sharp. The choir find it easier to sing in the key of the sympathetic note. Further, in many large churches where there is no sympathetic note it is still often found that tenor A is the best note to recite on or to intone the service on. Nothing is known about the reason for this, or how to design a building to get it.

Multi-purpose Halls

These have been dealt with in the preceding chapter, and here we will only briefly consider their use for music. As has been said, it is more important to have the correct reverberation for speech rather than for music in such rooms, unless a speech-reinforcement system specifically designed for rather reverberant conditions is relied on.

The orchestra should not be confined within curtains behind the proscenium arch but instead should be brought as far forward as possible. Thus an apron stage is a big advantage. Reflecting surfaces to help the orchestra hear themselves are desirable even if they are only temporary structures of plywood. One of the worst weaknesses will be the poor hearing quality at the rear of the hall due to the flat floor and stage platform. It is therefore most advantageous if an overhead reflector is used or the ceiling is shaped to reflect sound towards the rear. Such a reflector could take the form of a proscenium splay.

Rehearsal Rooms

By this we mean rooms for orchestras or choirs to rehearse in and associated with the particular concert-hall or other building. Such rooms should simulate the conditions of the auditorium proper as far as possible. Thus for a concert-hall the platform in the rehearsal room should be the same as the one in the auditorium and so should the nearby reflecting surfaces. It will not be possible to copy exactly the acoustical conditions but the reverberation time should be about the same as the full auditorium, and this will usually involve making the wall the orchestra faces at least partly absorbent. This has the added advantage that the orchestra will not get the reflections from this wall that they might like to have but which would not occur in the auditorium.

MUSIC ROOMS

By this we mean rooms used for music teaching and practice. The acoustical requirements do not appear to be very exacting and a reverberation time of between 0·5 seconds for the smaller rooms and 1 second for the larger rooms will probably be suitable. It may be desirable to make the reverberation adjustable to suit individual preferences, e.g. by having curtains which can be drawn across one or more walls.

If the rooms are of solid construction (e.g. brick walls with concrete floor and plaster or solid concrete ceilings) then the reverberation time will be much too long, particularly at low frequencies. This will call for the use of absorbents, and for the low-frequency absorption wood panels on an air-space would be suitable.

It is desirable but not essential that some diffusion should be deliberately introduced, e.g. by modelling of the wall surfaces.

5

The Design of Studios

The overall planning and design of broadcasting and television studios is a complex subject much involved in technology and still undergoing development. The architectural aspect is largely affected, one might almost say governed, by technical requirements, among which those of acoustics and sound insulation are prominent. In fact, the 'sound' requirements have such a strong influence on the general planning of broadcasting buildings that we cannot very well discuss one without the other. We shall therefore devote some space to planning and purely architectural considerations.

The acoustic design of studios is a specialised aspect of room acoustics. Although the general principles to be applied are the same as in other rooms for speech or music, there are certain unique requirements and the design is generally worked out in greater detail. Adequate sound insulation both from the interference of outside noise and between various rooms comprising studios and other technical and non-technical areas must be provided. This is dealt with in Chapter 8.

The main reasons for a different approach to the acoustic design of studios from that of rooms for normal audition are:

(a) that the sound is to be conveyed to the listener by a single channel of information, i.e. there is no provision for the transmission of the two slightly differing sounds which a listener to 'live' material normally receives, one in each of his ears. This neglects the possibilities of future 'stereophonic' sound transmission;

(b) that the acoustics of the room in which the sound will eventually be reproduced, as well as those of the originating studio, have some influence on the sound finally heard.

Normal (binaural) hearing gives the listener the power to

select some of the total sound and subconsciously partly to reject or ignore the rest. This faculty is mainly but not entirely due to the 'direction-finding' properties which are developed when two 'receivers' (ears) are used. This is possible because the direct sound rays from the source to the receiver are highly directional whereas the reflected (reverberant) sound is non-directional. It would therefore seem that the reverberation of a room when heard monaurally (i.e. through a single microphone to reproducer chain) should appear greater than when listened to in the room. This is borne out in fact and for this reason studios are made slightly 'deader' (i.e. have a lower reverberation time) than rooms for direct listening.

Because the reverberant part of the sound has in general a degrading effect on definition or clarity, although some reverberation is needed to give 'character' to the sound, it is important that its amount be closely controlled. It is common experience that the balance between reverberant and direct sound can vary sufficiently in an auditorium for it to be plainly noticed by normal listeners. This effect is even more marked in microphone reception because the microphone is unable to discriminate between the two parts of the sound. Although it may be almost always possible to find a suitable position for the microphone where the relation between the two parts of the sound is ideal by a process of trial and error, it is desirable to reduce this difficulty to a minimum. This is particularly true when, as is common, several microphones are employed at the same time, the outputs of which are combined to produce the final result. It is best, therefore, if the reverberant sound field in a studio is as far as possible uniform or diffuse and unless special areas of varied acoustical properties are needed, as in drama studios, acoustic design should aim at this result so that microphones can confidently be placed in almost any physically convenient position.

The effect of the acoustics of the room where the sound is eventually received or reproduced is also to add to the effective reverberation time. For example if the listening room has an average reverberation time of 0·5 second the total reverberation time can never be less than this even if that of the studio is less. When the studio has a reverberation time of 0·5 second the total reverberation time is 0·61 second. For higher values of studio reverberation the effect of the listening room becomes less and less until at a studio reverberation time of about 1½ seconds the total time is practically the same.

BROADCASTING STUDIOS

Planning—General Arrangement

In any but the smallest of broadcasting centres, there will be a group of studios designed to deal with a variety of programme material. The largest type of studio for sound broadcasting is the music studio for symphony orchestras and choirs. This is in effect a concert-hall, and is very often equipped with seating for an audience of at least one hundred and fifty and sometimes considerably more. The planning of this type of studio is much influenced by the facilities to be provided for the audience. These may be very elaborate and can include, as well as the minimal provision of public access and toilet facilities, cloak-rooms, foyers and associated restaurant, bars, etc. Smaller music studios may be required, suited to instrumental combinations varying from a medium-sized orchestra or brass band down to a trio or quartet or very small dance band. Comments on the desirable size for music studios are given below.

Another type of studio in which seating for a public audience is commonly provided is that for variety shows. This is virtually a theatre and many old theatres and cinemas have been pressed into service for the purpose. When such a studio is designed *de novo* it is not usual to provide for more than a small audience, say up to 400 persons. This type of studio will probably tend in future to merge with or be supplanted by the television variety type of studio for obvious reasons.

All of these studios require a control cubicle (of about 20 m²) with an observation window giving a wide view of the studio floor. In the largest studios the cubicle may be on a floor level higher than that of the studio, which would be two normal storeys in height. Rapid and easy access must be provided between the cubicle and studio floor. There may also be a recording room (adjoining the control cubicle and with vision both into it and the studio), a lighting control room (in theatre studios) and a narrator's studio. This last is used not so much for normal announcements, which are quite commonly done from the main studio, but for a commentary or 'narration' which may form part of the programme build-up. The smaller sizes of music studios are sometimes used as 'general-purpose' studios, in which speech may form a considerable part of the subject matter. If the reverberation time is too long for the purpose, acoustic screens (see p. 127) can be used to form a local area of acoustic condition more suited to the purpose.

Drama studios have undergone a change since the early days of broadcasting. It was once customary to use a group of studios, each with its own distinctive acoustic qualities, and a separate room in which all the 'effects' were produced, whether by apparatus or from recordings. The producer then had to mix the outputs from all sources on a 'dramatic control' panel to produce his final pastiche. This method of working, which had many difficulties, is now largely supplanted, at least in Britain, and the current method is to employ one large studio in which the whole of the cast can assemble, including effects operators and even one or two muscians for providing incidental items of a not too ambitious nature. A floor area of at least 60 m² is required with provision for division into two parts of contrasting acoustics—a 'live' end and a 'dead' end. The 'division' can be made by thick double curtains spaced about 1 m apart or by a sliding folding partition or by a combination of these. A control cubicle should be sited against the longer wall with the observation window arranged so that a view into each of the two halves of the studio is obtained. In addition a recording room may be required (with observation window into control room) or this function may be combined with the control room, which will then need to be larger. It is an advantage if the floor level of the cubicle is a few feet higher than that of the studio.

The studio itself is often usefully supplemented by one or two smaller rooms which are acoustically treated very dead and very live. These may either be small enclosures built within the main studio, if it is large enough, or be small adjoining rooms. The very dead room need be no more than about 4 m² in area, but the very live room is better made at least 8 m² in area. Both rooms or enclosures should, if possible, be arranged so that direct visual cueing is practical from the control cubicle. An example of this type of drama suite is illustrated in Fig. 34.

Talks studios include a number of types varying from a small room for a single speaker, such as a news reader, to a fairly large room suitable for a discussion programme with up to six or more participants. Each talks studio is provided with its own control cubicle, except where it forms a 'suite' (as for example the narrator's studio adjoining a large music studio), when it may share the cubicle with the other studio. There are two special types of talks studio suites which are discussed below.

Mixer suites consist of two rooms in which provision is made for building a programme from material originating from several sources, e.g. outside broadcasts, other studios or recorded material. One of the rooms is an announcer's or commentator's studio and is acoustically treated as a normal talks studio and the other as a control room. Occasionally the second room also has a microphone for the origination of programme speech as

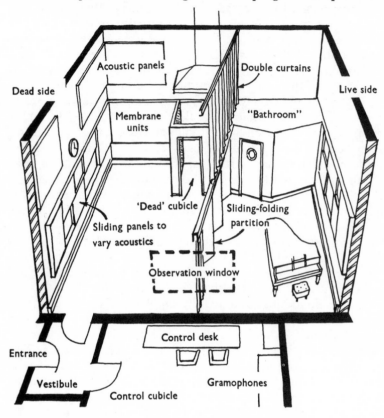

FIG. 34. The Layout of a Drama Studio

distinct from the usual 'talk back' microphone for cueing and instruction purposes, in which case this room also must be treated as a talks studio. Record-playing apparatus is housed in the control room (except in continuity suites). The rooms must have visual communication by observation windows.

Continuity suites are a particularised form of mixer suite. The continuity suite is the last but one link in the programme chain between the studio centre and the land line going out to the transmitter (see Fig. 35). The last link is the main control room but this does not present a particular acoustic problem. The

suite consists of two rooms, a studio in which an announcer sits
at a desk which is equipped with record-playing facilities as well
as the normal microphone, and a control cubicle. This last is

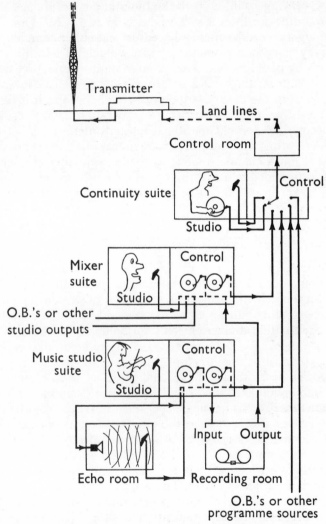

Fig. 35. Key Diagram of the Linking of Studios in a Broadcasting Studio Centre including Echo Room

similar acoustically to a normal control cubicle but is provided
with a switching apparatus to enable the engineer to select the
programme from a number of incoming line sources, or from
the adjoining announcer's studio. Occasionally continuity suites

consist of three rooms, a studio, a mixer room and a control cubicle. In this case programme microphones will be installed in the studio and mixer rooms (treated as talks studios) and record-playing facilities in the studio and control rooms.

Recording facilities may be housed in two ways. One is by planning a recording room adjacent to and either completely or partially separated from the control cubicle of a studio. The other is by planning one or more large multi-channel recording rooms. All recording rooms are acoustically treated as 'listening' rooms but in the case of large multi-channel rooms it is best to provide very dead acoustics (reverberation time of about 0·3 second) to prevent too much mutual interference between the monitoring loudspeakers. Strict quality checking is not done in these multi-channel rooms; a smaller adjoining or enclosed room is provided which is treated as a good listening room, the loudspeaker in which can be connected to any of the recording channels.

OPTIMUM SIZE AND SHAPE

The ideal size for a studio is influenced both by acoustical requirements and the uses to which it is to be put. In the talks studio group, if the acoustical properties of a room are to be allowed to contribute anything to the sound output, it is essential that the smallest dimension shall not be less than about 2·4 m. When dimensions in the range 2·4–3 m are involved, considerable care must be used to ensure adequate low-frequency sound absorption so as to avoid the bad effects of wide room mode spacing inevitable in such small rooms (see Chapter 2, p. 48).

Sometimes rooms for single speakers are required to be even smaller than this, and then the only satisfactory solution is to keep the rooms as acoustically dead as possible at all sound frequencies and to employ a 'lip' microphone or a close speaking technique which will ensure that the microphone output is predominantly direct sound. This approach to the problem will give the very 'direct' kind of speech sound commonly used for commentaries both in broadcast and film technique. If this acoustic effect is required, and there is probably much justification for it, particularly when the speech is to be merged with some form of background sound effect such as soft music, cheering crowds or the like, then quite a small cubicle, say 2 m by 3 m, provided it is acoustically dead down to the low-frequencies, is perfectly adequate. Continuity and discussion studio plan sizes are usually dictated by the physical space required for the occupants, furniture and apparatus. For talks studios generally, a ceiling height between 2·4 m and 3·7 m (depending on the plan area) is appropriate.

The size for drama studios is governed largely by space requirements. It is important to observe that an 'action' area of about 2 m radius is needed round the microphone. This space will be used by actors in obtaining some perspective effects and as grouping space when several of them have to speak simultaneously or in quick succession. The height of drama studios should be between 3 m and 6 m depending on their area.

Where it is required to provide for so-called 'audience participation' a certain floor area must be set aside for the audience seating or they may be accommodated in a balcony. To maintain the proper performer-to-audience relationship it is necessary to keep the audience a certain minimum distance away from the action area, nor must this area be encroached upon or reduced for the sake of audiences.

For music studios a size can be chosen based on the number of performers. The B.B.C. have used the following table as a general guide:

TABLE III

No. of Performers	Minimum Studio Volume
4	42 m³
8	110 ,,
16	340 ,,
32	850 ,,
64	2300 ,,
128	6200 ,,

These suggested sizes are based on the requirements for 'classical' music and instruments. It is well known, however, that artistic fashions in serious and to a greater degree in light music vary among people and are always changing and that the acoustical background required to be provided by the studios can be strongly influenced by these artistic preferences. However, to enable musicians of whatever heights of brow to develop adequate musical tone, the minimum volumes quoted above are strongly recommended.

In deciding the necessary floor area it is useful to remember that instrumentalists will require at least 0·8 m² of floor space each, on an average, and that an area at least twice as large as that taken up by the orchestra must be allowed for microphone placing. To this total should be added about another 50% for the smaller studios or up to 150% for larger studios, for circulation space. Moreover, although complaints from artistes and producers that a studio is too small have frequently been heard, complaints of a studio being too large are much more rare.

Summarising, studios should at least be large enough comfortably to accommodate all the possible physical contents but may require to be even larger for acoustical reasons.

There is probably no optimum or best shape for a studio. For rectangular rooms certain preferred ratios of the three dimensions are often quoted, such as 1:1·25:1·6. These ratios are based on formulae which give the most even distribution of room modes in the lower part of the sound spectrum. If studios were rooms having little or no absorbents these proposed ratios would very likely be the best ones to choose, but since studios in practice contain a large amount of absorption, the choice of such exact relationships in the sizes becomes unimportant. However, it is advisable to avoid making studios, particularly in the smaller sizes, with their dimensions related by simple numbers such as 1:2:3, etc. The worst possible shape would be a cube, for example, and then a double cube. The departure from such ratios need not be more than 0·2 m to 0·3 m. It is usually better both acoustically and architecturally to avoid ratios of any pair of the room dimensions greatly exceeding 1:2·5. This may lead to the minimum cubic capacity of a studio giving a room of more than usual height. For music studios this is not a disadvantage. Rooms 4·6 m to 6 m high and of only 30 m² floor area often give very good musical tone, whereas a studio of the same volume but only 3 m high may not be so successful.

Rooms with non-parallel boundary surfaces are often advocated, in which a whole wall or walls or the ceiling is deliberately canted out of parallel with the remainder of the structure. A fan-shaped plan is a typical example. Occasionally this type of shape can result in prolongation of reverberation in the form of a 'flutter' echo due to fortuitous repeating patterns of sound 'rays'. The illustration Fig. 36 shows a typical example. Apart from this disadvantage, wide departure from rectangular shape may disturb the random distribution of sound energy. This has two drawbacks: first that it is undesirable in itself, and secondly that it is probable that because the sound distributes itself less randomly over all the boundary surfaces, the effect of absorbents placed on the surfaces will be less accurately predictable. The reverberation formulae used in calculation have as a premise perfect random distribution, and therefore rooms where this premise is not realised are more likely to have a different actual reverberation time from that calculated during design.

We do not suggest that it is essential for all studios to be rectangular; if planning expediency requires a studio with non-parallel walls it can be accepted subject to the foregoing reservations.

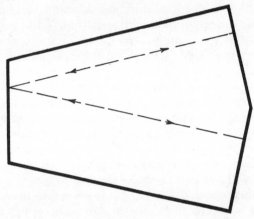

FIG. 36. Development of 'Flutter' Echo in a
Studio with Non-Parallel Walls

Apart from major diversions from a plain rectangular shape, many possibilities of irregularity in wall, ceiling and even floor shapes exist. This irregularity may take the form of extensive breaking up of surfaces by projections or recesses of fairly large dimensions or may be merely a comparatively slight modelling in the form of applied decorative features. In either event the effect will be to produce more 'scattering' of sound and so to increase diffusion, which is generally desirable. Traditional architectural detail such as columns, pilasters, friezes, cornices and so on, are admirable sound diffusers, but because such detail is rarely used in contemporary architecture, arbitrary shapes are sometimes employed. Some question exists as to whether rectangular, convex curved (polycylindrical), triangular or some other shaped projections are most effective. The measured acoustical difference between the various shapes, assuming they are of roughly the same dimensions, is so slight that the decision as to which to use can probably remain a purely architectural one.

Diffusion also occurs at the boundaries of any special acoustic treatment; that is, where a surface changes from one type of absorbing characteristic to another, irrespective of any change in the plane of the surface. This is the so-called 'edge-effect' of absorbents (see p. 50). If acoustic absorbents are distributed in patches over the available surfaces as is recommended, it follows that a fair amount of diffusion will be provided, but since, with few exceptions, most absorbents are highly efficient only in the middle to high frequencies, the best diffusion will be obtained only in these regions of the sound spectrum. This probably does not much matter in studios used solely for speech because the sound energy produced by the voice at frequencies below 200 Hz is very small.

In studios for music there may be some lack of diffusion of the low-frequency sounds and it is advisable to obviate this by the use of diffusing projections. To be effective the minimum depth of such projections needs to be about one-seventh of a wavelength, for example up to about 0·6 m for sounds down to 80 Hz. As music studios will generally be rather large rooms, projections of this size will not be too coarse in scale to give a good architectural appearance. However, if it is desired to maintain a flat appearance to the room surfaces it will be necessary to provide the irregular diffusing and sound-absorbing elements and then to obscure them with some flat screen-like covers which must, of course, be sufficiently 'transparent' to the sound waves.

Solid free-standing objects such as furniture, fittings, columns or light fittings also provide diffusion and often no further provision is made. Studios are, however, often rather bare of such objects compared with, say, domestic rooms and can easily suffer from lack of diffusion. If a few pieces of fixed or movable furniture of appropriate character—such as banquettes, occasional tables, etc., are introduced into talks or discussion studios they perform the useful functions of providing sound diffusion and taking away the severity of the appearance of the room.

It is dangerous to create or admit recesses or adjuncts to the room which are of such size and character that they might have individual acoustic properties of their own apart and different from those of the main body of the studio. If, for example, the reverberation time of the adjunct is longer than that of the main body of the room, sound will be fed back from the adjunct after the sound level has fallen to a lower value in the rest of the room. On the other hand if the adjunct is deader than the main part of the room this danger is avoided.

PRACTICAL ACOUSTIC DESIGN

Details of the practical sound insulation design of studios will be found in Chapter 8, p. 199. The acoustic design of studios follows the same general lines as those described in Chapter 3, p. 72, except that it is recommended to use the Eyring formula for reverberation time instead of the Sabine formula (see Chapter 2, p. 53) and to calculate values for each octave from 62 Hz up to 4000 Hz, with an average value to agree with the optimum value given from the graph, Fig. 37. It should be

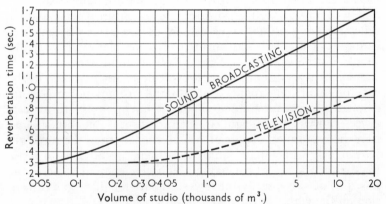

FIG. 37. Recommended Reverberation Time for Broadcast and Television Studios

remembered that the calculated values for the frequencies 62 and 125 Hz are liable to be rather more inaccurate than those for other frequencies because the low-frequency sound is rarely so completely diffuse in the room and because unpredictable variations in absorption coefficient are more likely at these frequencies. A nomogram of values of $-\log_e (1 - \bar{\alpha})$ against values of $\bar{\alpha}$ is given in Fig. 14.

Having determined the required amount of absorption at each frequency it remains to choose absorbent treatments which will, in conjunction with the absorption of any untreated surfaces, give these values of total absorption. It is best to ignore the absorption of contents of the studio (personnel, furniture, etc.), except in the case of studios with permanent provision for audience when the seats and audience may represent a fairly large proportion of the total absorbing material, or of those with extensive rostra for orchestral players. These last provide a certain amount of mainly low-frequency absorption. Coefficients for typical timber rostra are given in Appendix A.

In deciding on the arrangement of the absorbents, apart from the obvious consideration of ensuring that the treatment is suited to its physical location by using for example robust, hard-wearing materials on the lower part of the walls, there are two important factors to be borne in mind. First, that the absorbents are distributed so that roughly the same amount of absorption (at all frequencies) is applied to each of the three pairs of opposite surfaces. Secondly that the various types of treatment used are divided up into a number of separate patches or areas and well mixed. It is also desirable to ensure that extensive reflective areas do not come exactly opposite one another. For example, if a studio has an area of plain plastered wall, this should not be exactly opposite the observation window.

A slight reservation to the above rules must be made in the case of small long rooms—for example, rooms not exceeding 4 m in their smallest dimension and having a length at least twice their width or height. In such rooms it is sometimes found that the room modes associated with the length of the room are particularly difficult to damp. It is therefore advised to use a low-frequency absorbent having high efficiency at some of the low, but not necessarily the lowest, modes for this room dimension on the whole or the major part of one end wall.

Floors in talks, discussion and drama studios are best provided with carpet or some floor covering which is equally quiet underfoot. Carpet provides appreciable sound absorption, an advantage not shared by soft-based rubber or other smooth surfaces. In drama studios some part of the floor surface is often used to provide footstep effects. Wood strip on battens, wood blocks on concrete, smooth concrete and a shallow trough filled with gravel are the usual surfaces provided. Similarly a short flight of stairs (sometimes divided down the centre with two different surface finishes) and various effects doors and windows with latches, knockers, etc., can either be incorporated into the design of the studio or may be required as portable units. The floors in music studios, at least the areas which will be used to seat the orchestra, should be either wood strip or wood block. General circulation areas of music studios, for example any audience area, can be carpeted, but a reflective surface, preferably of wood, is properly regarded by musicians to be essential as a playing area. It should be noted that if a board and batten floor is used where grand pianos are to be installed, the batten spacing must be very close and the boards thick, hard and strong enough to resist the considerable point loads occasioned by these instruments when mounted on castors as they usually are.

The photograph Plate V shows a small talks and discussion studio. To illustrate the operations of acoustic design this studio is fully described in Table IV, all the calculations are shown and the calculated and final measured reverberation times are given.

The illustration Plate VI shows a general-purpose music studio. In this studio no membrane low-frequency absorbers or porous medium high-frequency absorbers are used. The entire acoustic treatment is by Helmholtz resonator type absorbers. The rows of rectangular projections seen on the walls are groups of low-frequency resonators formed of fibrous plaster. The medium- and high-frequency absorption is by multiple resonators formed by perforated plasterboard with a thin layer of scrim cloth laid on the back. The plasterboards are variously spaced from the solid walls or ceiling so as to spread the resonance peaks of absorption over the necessary wide band of frequencies.

DRAMA STUDIOS

The need to provide a variety of acoustical environments in one studio can be met by dividing the room into two parts by means of curtains or folding partitions and treating one part as a comparatively 'live' area and one as a 'dead' area. In designing the two parts of the studio an average reverberation time of less than the optimum value (given in Fig. 37) should be chosen for the 'dead' part and a value at least as high as the optimum for the 'live' part. The calculations can be made treating the volume of each part of the studio as a separate room and allowing absorption for the dividing curtain calculated from the coefficients given in Appendix A. Alternatively at frequencies below 500 Hz where the curtains have practically no surface absorption and transmit most of the sound falling on them into the other part of the room an amount equivalent to the total absorption on the far side of the curtain can be included to represent the

TABLE IV

REVERBERATION TIME CALCULATION FOR A TALKS STUDIO (SEE PLATE V) BY THE EYRING FORMULA

Size: 4·73 m long × 4·43 m wide × 3·2 m high. Volume: 66·8 m³ Total surface area (S): 101 m²

Optimum reverberation time (from Fig. 37): 0·3 sec. approx.

Frequency (Hz), Absorption Coefficient and Absorption units

Surface	Area m²	62 α		125 α		250 α		500 α		1000 α		2000 α		4000 α	
Floor, pile carpet on wood boards	21	0·15	3	0·2	4	0·25	5	0·5	10	0·5	10	0·6	13	0·65	14
Ceiling, lath and plaster	14·5	0·1	1	0·3	4	0·15	2	0·1	1	0·05	1	0·04	1	0·05	1
Membrane units (roofing felt) 250 mm deep	10	0·9	9	0·6	6	0·3	3	0·2	2	0·1	2	0·1	1	0·1	1
Plywood panels on battens with glass wool in cavities	37	0·2	7	0·3	11	0·2	7	0·15	6	0·1	4	0·1	4	0·1	4
Fabric over 25 mm glass wool	5·6	0·08	nil	0·15	1	0·35	2	0·7	4	0·85	5	0·9	5	0·9	5
Plate-glass window	1·4	0·1	nil	0·1	nil	0·05	nil	0·04	nil	0·03	nil	0·02	nil	0·02	nil
Wood door	1·7	0·1	nil	0·15	nil	0·2	nil	0·1	nil	0·1	nil	0·1	nil	0·1	nil
Perforated plywood (10%) over glass wool	9·5	0·05	nil	0·15	1	0·35	3	0·75	7	0·85	8	0·75	7	0·4	4
Totals	101		20		27		22		30		29		31		29
ᾱ			0·2		0·27		0·22		0·3		0·29		0·31		0·29
$\log_e(1-\bar{\alpha})$			0·22		0·32		0·25		0·36		0·34		0·36		0·34
Sabins			22		32		25		36		34		36		34
Air absorption	66·8 (m³)		—		—		—		—		nil		nil		1
Total absorption			22		32		25		36		34		36		35
T. Eyr calculated			0·48		0·33		0·43		0·3		0·31		0·3		0·3
Measured reverberation time			0·35		0·37		0·33		0·28		0·27		0·28		0·29

absorption of the curtain 'wall'. This method seems to give a closer estimate of the true reverberation time. No attempt should be made to provide any high degree of insulation between the two parts which will be used for drama productions concurrently rather than simultaneously, the cast of artistes moving freely about to the microphones in the various parts of the studio during the production, according to the acoustical effects required. A typical design for such a studio at one of the B.B.C. studio centres is shown in Fig. 34. Small 'extra dead' and 'extra live' adjuncts have been provided and a door is fitted to the live room to avoid the dangers of return sound mentioned above. When the dividing partition is folded back the whole room has characteristics which are again slightly different from the two

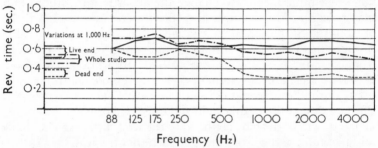

Fig. 38. Reverberation Time at Various Frequencies in a Drama Studio

parts, and in this way a wide choice of acoustical facilities is made available to the producers.

To widen the scope further, some variable elements are incorporated in the walls. These consist of hinged or sliding panels with absorbent surfaces (in the live end of the studio) or hard reflecting ones (in the dead end) which can easily be deployed so as to cover the normal wall surface which has the opposite type of acoustic character to that which the panels will provide. The net effect of these variable elements in the overall reverberation time is admittedly small (see Fig. 38), but owing to the fact that the panels are at normal microphone level and that this particular studio is rather small in plan so that microphone positions fairly near to walls are commonly used, the subjective effect is rather greater than the changes in reverberation time would indicate. The effects achieved, however, are only comparable to those which can be obtained by the use of acoustic screens. A typical example of these is shown in Plate VII. They are provided with reflecting and absorbent sides, a rotating centre panel so that the screen can be reversed without moving it bodily and an extending section to increase the total height. Three, four or more of these screens are used to create small local acoustical influence on quality, particularly of speech.

The alternative method of dividing this type of studio where more area is available is by means of two parallel curtains, spaced about 1 m apart. With this arrangement it is quite common for the space between the curtains to be used for special dead acoustic effects.

VARIABLE ACOUSTICS

Broadcasting organisations often demand the facility of being able, by simple means, to change the acoustics of a studio to a major degree. Many essays have been made at this. It will be obvious that if the acoustical change is to be large a considerable proportion of the total surface of the studio must be changed from a material of low to one of high absorption. Unless an elaborate mechanical system operable by remote control is to be used (with the attendant disadvantages of such systems, that they are difficult to maintain in good working order) the variable elements will need to be confined to the wall surfaces. The shape and planning of the room must be such that the wall areas available for this are as large a proportion of the total surface as possible. For example, in a room 5 m by 7 m by 3 m high the total wall area is less than a half of the total surface area, and this would therefore be an inappropriately shaped room in which to attempt this design.

Much ingenuity has been used in the past in designing surfaces with variable sound absorption, and no doubt it is a project which will give satisfaction to many designers in the future. Whether the effort and cost involved are justified by the improvement in usefulness of the studio is another matter. It is often reported that the variable elements, where they have been provided, are set to one particular condition and remain there for most of the working life of the studio. It is often simpler to make any required adjustments by the use of acoustic screens of the type mentioned above.

Listening rooms, that is control cubicles, recording rooms and any technical room where programmes may be listened to or 'monitored', should be acoustically treated to resemble as closely as possible the acoustic conditions in an average domestic living-room. In this way the programme engineers will have a good idea of the final sound emerging at the end of the broadcasting chain. The average reverberation time of domestic rooms is known to be between 0·4 and 0·5 second, and it is desirable to ensure that the reverberation time is substantially the same as this over the frequency range 100 to 4000 Hz.

TUNING

Some degree of preadjustment of the acoustic treatment (tuning) should always be provided. Acoustic calculations are even now at best only a good guide to the finally achieved conditions, and unless the practical design of treatment is such that simple modifications can be carried out before the studio is put into service, disappointment may result. For this reason exclusive use of types of treatment such as acoustic tiles which are not readily 'adjusted' is deprecated. Treatments consisting of a masking surface such as perforated or slotted, hardwearing, and decorative sheeting over porous, membrane, or Helmholtz resonator type absorbers are much to be preferred because of the ease with which the overall characteristics of such treatments can be changed. It is also strongly recommended that where possible, reverberation time measurements should be made at suitable intervals in the course of construction of the studios. This will enable the designer to see if the conditions are tending towards the desired final result.

TELEVISION STUDIOS

Because of the comparatively recent inception of television services the available experience on the best design for studios is relatively small. The operational conditions normally found in television studios had been approximately met in film studios, but even here the need for acoustical design became important only after the introduction of sound films. In the earliest days of television, as in those of sound films, the problems of obtaining good pictures were allowed to override the need for good accompanying sound, and this may be true to some extent even now.

When sound was first added to cinema film, the initial step taken in the acoustical design of studios was to make them very 'dead'. This was partly because the scene of much of the material of films was out of doors and sound with any reverberant background was so plainly out of harmony with outdoor pictures. Interior scenes with 'outdoor' acoustics did not seem so ill matched. However, when 'musicals' became fashionable it was quickly realised that a completely dead studio is the worst possible environment for the playing and recording of orchestras. This difficulty was met by the provision of special 'sound stages' (in effect broadcast studios) in which the orchestras play while the conductor watches a projection of the film and the music is 'dubbed' on to the sound track. Generally speaking no filming is done in sound stages nor are orchestras introduced into the film shooting studios.

The production of television programmes at the moment requires different techniques, the main one which concerns studio design being that the musical part of the programme must be integrated much more closely with the visual part. Some attempts were made to employ the film method in the B.B.C. but they were largely unsuccessful. Music recorded in a dead studio can, of course, be made to sound better at the receiving end by the use of artificial echo, i.e. by adding some reverberant sound artificially to the music, but this does nothing to provide the musicians themselves with tolerable acoustic conditions which are essential if they are to produce their best work.

It therefore became necessary to find a degree of acoustical liveness which is not so great as to make it difficult to obtain a good sound match for outdoor scenes played in the studio but at the same time is enough to provide acceptable conditions for musicians and artistes. This was done experimentally in B.B.C. studios by progressively 'livening' dead studios until satisfactory values of average reverberation time were obtained. It appears, however, that the acoustics of television studios are not so critical as those in broadcasting studios.

PLANNING

Generally, acoustical requirements are not governing factors in the size or shape of television studios. There appear to be four main types of studio to be considered, namely:

(a) general-purpose studios for all types of programme with no, or only makeshift, provision for 'audience partici- pation';

(b) studios, usually of a 'theatre' type with permanent audi- ence seating;

(c) small interview and 'announcer' studios;

(d) 'dubbing' suites.

Studios of types (*a*) and (*b*) are commonly very large. Volumes of up to 10,000 m³ are known. A maximum of clear floor area is required and usually a minimum height over the working part of the studio of 8 m must be provided to allow for the elaborate lighting grid and scenery flying gear (see Plate VIII). Each studio is provided with the following ancillary technical rooms:

(*a*) vision control room;
(*b*) sound control room;
(*c*) lighting control room;
(*d*) camera control and apparatus room or rooms.

In addition there are make-up, quick change, property and other store rooms which must be very close to the studio. Of the above list (*a*), (*b*) and (*c*) are commonly arranged in a suite and must have good visual connection with the studio. They are therefore usually sited one storey higher than the studio floor, with the observation windows preferably in the centre of one of the studio walls. The rooms (*d*) do not require visual connection with the studio but must be reasonably close to the control rooms to avoid long runs of technical wiring. The necessary size and planning of all these rooms is largely dictated by the technical apparatus which they have to house.

These studios must be provided with very flexible arrangements for the erection and lighting of sets. Usually a lighting gallery is provided at about 4 m above floor level all round the studio and invariably a system of catwalks or even a complete gridded floor over the whole studio is used to give access for adjustment to the electric lighting equipment. Powerful and elaborate ventilation plant is required to reduce the heat created by the high-powered lights and sometimes a refrigerated air-supply system is regarded as necessary.

Many old theatres have been converted into audience participation television studios, and the stage facilities (scenery fly tower, lighting, etc.) have been adapted to television use. The major disadvantage of these adaptations is the smallness of the stage. In many cases a large apron stage has been added to overcome this but even so the proscenium arch is often an embarrassment. Those theatres with very large stages usually have large auditoria but the participating audience for television productions rarely exceeds a few hundred.

The small interview or announcer's studio with facilities for vision output requires to be considerably larger than comparable sound studios (see Plate IX). A floor area of 60 m² can be taken as a minimum and the clear ceiling height should be at least 4 m. One control cubicle is usually considered sufficient, but this may form with two or three other rooms a continuity suite. The group might consist of:

(a) presentation studio;
(b) control cubicle;
(c) central control room—vision;
(d) central control room—sound;
(e) sound and vision quality check room.

If possible rooms (a), (b), (c) and (d) should be visually linked by observation windows.

Either of two methods of planning dubbing suites may be used. In one the projection theatre itself is used as a sound studio. The planning of the theatre follows the usual lines with a projection booth, but a control cubicle with observation window and sometimes a recording room are also provided. The alternative method of planning is to have a projection theatre with a small commentator's booth adjoining equipped with an observation window giving a good unobstructed view of the screen. In this case a control cubicle with observation windows into both projection theatre and commentator's booth can be used, or alternatively the control of sound can be done in the projection theatre itself.

Acoustic Design

For all types of television studios the average reverberation time should be adjusted to a value depending on the volume of the room, given in Fig. 37. The reverberation time at frequencies between 100 Hz and 4000 Hz should be as nearly as possible constant. A rise in the low-frequency reverberation to not exceeding 1·5 times the average value will matter less in the very large studios than in small presentation or sound-dubbing studios. Similarly, in the largest studios some drop in the reverberation time at very high frequencies is inevitable because of air absorption and must be accepted.

In small commentators' booths a very short reverberation time should be provided, say 0·25 to 0·3 second at all frequencies down to 100 Hz.

An important acoustical aspect of the design of studios for audience participation is the need to provide an adequate sound coverage for the audience. Quite apart from the fact that many light-entertainment artistes employ a microphone and the distinctive acoustical quality resulting from its use as part of their technique, there is the physical impossibility of obtaining enough sound energy from the unamplified human voice adequately to fill a large and comparatively dead room. It is therefore essential to use a sound-reinforcement system in these studios. The need to avoid amplified sound getting back into the microphone and so giving an unwanted reverberant character to the broadcast sound or even, in extreme cases, setting up 'howl-round' can best be satisfied by the use of highly directional loudspeakers such as the line or column type and low reverberation time. Very often when old theatres are taken over for television, the reverberation time (especially in view of the fractional audience present) is often too long and extra acoustic treatment has to be introduced. Further details on the correct design of sound-reinforcement systems are given in Chapter 6.

The acoustic treatment of projection theatres, sound-control rooms and sound-quality checking rooms should be designed to give good listening room conditions (i.e. average of 0·4 to 0·5 second). In vision control rooms, although monitoring loudspeakers are in constant use, the primary consideration is not the judgement of sound quality. The production technique calls for constant verbal communication between the producer and his assistants and (over intercommunicating telephone or radiotelephone) with camera operators, microphone boom operators and studio managers on the floor of the studio. All of these sounds must be kept to some extent mutually non-interfering and this can only be done by providing acoustically very dead conditions. An average reverberation time of not more than 0·25 second has been found acceptable.

Lighting-control rooms, apparatus rooms, central control rooms and any other area where purely nominal sound monitoring is done require a minimum of acoustic treatment such as would be provided by an acoustic tile ceiling.

Details of the practical sound-insulation design of studios will be found in Chapter 8, p. 199. The practical details of acoustic treatment can follow the same general lines as those for sound broadcast studios, although the larger general-purpose television studios are usually treated in a rather more 'utilitarian' manner. Mineral-wool mattresses in cotton scrim and chicken-wire mesh fastened to or between battens on walls and ceiling are commonly employed, although there is a need to mingle with this treatment some units which will provide good low-frequency absorption such as membrane units or plain panels of plasterboard, hardboard or plywood. The rock-wool type of treatment is a legacy from film studios and is not by any means the only suitable method. An overall treatment using wood-wool slabs (which can be decorated by spray painting), with some panels plastered to give low-frequency absorption, has been successfully employed.

ARTIFICIAL ECHO

The provision of a facility for adding to the sound output of a studio some additional reverberation or 'echo' is a very common requirement for sound and particularly television broadcast studios. This can be done in two ways. The first is by the use of an 'echo machine', which is a piece of electronic apparatus, often using recording techniques, and demanding only a small floor space in a technical area for its accommodation. This method will not be described in detail.

The second and older method is by the use of an echo room. Fig. 35 gives a key diagram of the arrangement used and some details of the acoustical requirements for such a room follow. The echo room must be as large as possible, and since it represents a complete loss of space for any other purpose, economic considerations often govern the size. Content of 60 m³ can be taken as a minimum. The shape should be approximately rectangular and dimensions related to one another by simple numbers should be carefully avoided. In fact, if there is any justification for the use of 'preferred ratios' (see above), then it is most likely to apply to echo rooms. Rooms with non-parallel walls are sometimes advocated but it is not thought that they possess any distinct advantages. A departure from rectangularity in the form of bold but not excessively numerous projections or recesses is not a disadvantage and echo rooms can therefore be planned in spaces that might otherwise be difficult to find a use for. The treatment of walls, floor and ceiling should be as hard and solid as possible. For walls, brickwork, finished in hard plaster and gloss painted, is a suitable construction. The ceiling should also be in painted hard plaster, preferably on a concrete soffit, but a good thickness (say 19 mm) of plaster on lathed joists should not introduce too much absorption provided that the area is not more than about one-sixth of the total surface area of the room, and the remaining surfaces are very solid. The floor should be of smooth concrete or a screed finish on concrete and can with advantage be sealed with a floor paint. Board on batten or joist floors should be avoided. Echo rooms are often conveniently located in basements where retaining walls provide very good reflection and intruding noise is low. It is important that the overall insulation of these rooms is good, up to studio standards both from the point of view of keeping intruding noise out of the microphone circuits and of protecting the surrounding rooms from the high noise levels produced by the loudspeaker in the echo room.

Where two echo rooms are planned adjacent to one another, a minimum average sound insulation between the rooms of 65 dB (insulation slope not more than 5 dB per octave) is necessary to avoid mutual interference between them.

There is no optimum value of reverberation time for echo rooms, the aim being to obtain as long a time as possible. For a room of about 85 m^3 volume, constructed in the manner described, an average reverberation time of between 4 and 8 seconds can be obtained. Occasionally the reverberation is excessively long in the low or low-and-medium frequencies, for example, when the surrounding walls and ceiling are exceptionally thick and solid, and it may then be necessary to insert some absorption into the room to level the curve off. Membrane units can be used for this purpose if the absorption is needed below about 250 Hz. On the other hand, if a more general lowering of reverberation time is required or more particularly if the room appears to have prominent standing waves of certain frequencies, causing rather coloured echo characteristics, a number of diffusing elements in the form of large flat baffles can be introduced with advantage. These baffles can be of thick plywood, plasterboard or chipboard, about 1·2 m square or a little larger, and should be set up at random positions and angles in the room. The surfaces should be gloss-painted. Alternatively a number of free-standing solids such as cylindrical columns of about 0·3 m in diameter and of varying height will produce the same diffusing effect.

6

The Design of High Quality
Speech-Reinforcement Systems

This chapter is confined to the design of permanent speech-reinforcement systems for use in buildings. The amplification of music is not considered, partly because there is such a variety of circumstances—ranging from dance music in a ball-room to a harpsichord in a concert-hall—that generalisation is not helpful, and partly because such installations are usually under the control of an expert, who will have his own decided views of what equipment to have and how to use it, often with successful results. Speech-reinforcement systems, on the other hand, have a deservedly bad reputation, and this is not because of shortcomings in the equipment but because the installers and users of such systems often have little understanding of the acoustical problems. It is the purpose of this chapter to explain these acoustical problems and to show how, with the help of recent developments, it is possible to install speech-reinforcement systems of much better quality than is often accepted.

Reinforcement systems have to be used for either or both of two main reasons. The first is that the unaided sounds would not be loud enough: because there is too much other noise, because some of the audience is too far away or because the speaker does not speak loudly enough. These factors are inter-related, because if a listener is close enough to the speaker, the intruding noise (within reason) is not important. Put the other way round, the level of the intruding noise will determine how far away from the speaker a listener can be before the speech he is trying to listen to is lost in the noise. There is of course a limit to this inter-relationship, because even if there were no noise at all, the speech sounds would eventually, as the listener moved further and further away from the speaker, get so quiet as to sound unnatural.

The second reason for reinforcement is that the acoustics of the building interfere with the unaided speech sounds, and a reinforcement system is required to mitigate bad acoustics.

We would define a good speech-reinforcement system as one with which all the audience hear the speech clearly, undistorted and at reasonable loudness. Further, in most cases the speech should appear to be coming from the human speaker and not from any loudspeaker, although there are exceptions—such as political meetings, where intelligibility is by far the most important requirement and illusion of reality is not important. Ideally, the audience should not be aware there is a loudspeaker system in use.

MICROPHONES

Although we do not intend to discuss apparatus in detail, it should be stressed that good quality is essential, and this applies particularly to the microphones. If a bad microphone delivers distorted speech signals to the rest of the system, nothing can be done subsequently to put this right.

Microphones are made either equally sensitive to sounds arriving at them from any direction (omni-directional), or more sensitive to sounds arriving from one or more particular directions. Directional microphones are nearly always used for reinforcement systems because they reduce the feed-back (i.e. the howling noise produced when the amplification round the circle microphone to amplifier to loudspeaker and to microphone again is greater than unity). Feed-back occurs at a particular frequency rather than at all frequencies at once because most loudspeakers vary in efficiency from frequency to frequency and because of the room acoustics. Further, poor-quality microphones will also have an irregular response. Thus feed-back will happen at some frequency at which the loudspeaker and/or microphone is most sensitive and at which the loudspeaker most easily excites a room resonance (eigentone) containing the microphone and loudspeaker. A directional microphone can be placed so that its least sensitive side faces the loudspeaker and thus the amplification of the system can be greater before feed-back occurs, but it will be seen later how this requirement can conflict with the requirements of a high-level system.

Directional microphones have either figure-of-eight directional characteristics (i.e. sensitive at front and back, insensitive at the sides) or cardioid characteristics (i.e. more sensitive at the front than at the back and sides). A good cardioid microphone

137

will have a front-to-back ratio of some 20 dB and this ratio should be maintained as near as possible at all frequencies. A ribbon microphone has a figure-of-eight characteristic, and is the most commonly used sort, partly because it is cheaper than the more complicated cardioid microphone, and partly because its sensitivity is at its maximum at the front, decreases steadily right down to zero exactly at right angles to the front and then increases to a maximum again at the back. The microphone can be placed so that it is exactly at right-angles to the loudspeaker, and assuming negligible reflections from the auditorium the feed-back can be reduced more than with a cardioid microphone where the frequency response is less at the sides and back than at the front but nowhere reduces to zero.

If the man speaking (we will use 'man speaking' rather than 'speaker' to avoid confusion with 'loudspeaker') is in a fixed position, e.g. at a rostrum, then microphone placement is not usually a serious problem. The microphone should be about 0·5 to 0·7 m from the mouth: if closer the speech gets frequency distorted, i.e. the balance of the natural voice is upset and with ribbon microphones the lower and middle frequencies are accentuated; if further away the microphone will pick up the reverberant sound. When listening to natural speech, the brain with the help of the directional properties of the two ears can discriminate to a great extent against the reverberation of the speech. Thus the speech will still sound good up to some distance from the man speaking, say, for example, up to 15 m in a hall with reasonable acoustics. The microphone has no such discriminating properties, and has only to be 1·3 or 1·6 m away from the man speaking for the speech picked up by it to sound very reverberant. The man must not of course move out of the sensitive area of a directional microphone. This is more important when a ribbon microphone is used because its sensitivity falls off to the side more rapidly than the cardioid. Nor must he move about too much, and this is because the loudness will vary a lot. For example, the man's mouth may normally be, say 0·6 m from the microphone; if he leans forward so that his mouth is only 0·3 m away the sound level at the microphone will go up 6 dB, and the sound from the loudspeakers will go up correspondingly. This will sound unnatural to the audience, because this slight forward movement under natural, i.e. unreinforced, speech conditions would make very little difference (less than 0·5 dB) in the loudness to a listener 6 m away, and even less difference to listeners further away.

The restriction of movement is sometimes a handicap to, for example, dramatic preachers, whose actions are thus limited, and is a more serious disadvantage in lecture theatres where the lecturer wishes to continue talking while he moves over to the blackboard or the screen. A lapel microphone, that is a very small microphone fixed in the lapel, can be used to give freedom of movement, but the lecturer must avoid getting tangled up in the lead.

Often, speech from several fixed positions has to be amplified. Each position will have to have its own microphone, and it is most desirable that only the one in use at any given moment should be 'live'. If the other microphones are left on, they pick up the reverberant sounds; this tends to confuse the speech.

The placing of microphones for amplifying stage performances is most difficult. Usually, several microphones are used, hidden from view and placed so as to cover the whole stage area. Because of the large and varying distance between man and microphone the amplified sound is reverberant and varies in loudness. This problem might be solved by 'super-directional' microphones aimed at the actors and which would thus discriminate against the reverberant sound more than ordinary directional microphones. However, this is still largely an unsolved problem.

LOUDSPEAKER PLACING

It is the placing of the loudspeakers that is most often responsible for the poor quality of speech-reinforcement systems. The behaviour of sound in rooms is described in Chapter 2, and similar considerations apply when dealing with loudspeakers. Briefly, for good intelligibility the direct sound, i.e. the sound travelling directly from the man speaking to the listener, should be of adequate intensity compared with the reverberant sound. The intensity of this direct sound gets less as the distance between the man speaking and the listener increases—by at least 6 dB every time the distance is doubled, and by more if the direct sound is passing at grazing incidence over the heads of the audience. On the other hand, the intensity of the reverberant sound is much the same over the whole audience area. Thus, a listener close to the man speaking hears the direct sound well above the reverberant sound and the intelligibility is good; a listener a long way away gets the direct sound at a much lower level and the intelligibility is worse. How much worse will depend on the reverberation time of the room and on the nature of the first few reflections of the sound, as discussed in Chapter 3. Further, the loudness must be kept at a reasonable level over the whole audience area, both for naturalness and to make sure that the speech sounds are well above any noise.

As an example we will consider a hall with a flat floor seating five hundred audience and where the reverberation time is about the optimum for speech. The distance between the man speaking and the first row of listeners we will take as being 3 m, and between the man speaking and the back row as 26 m. Obviously the intensity of the direct sound reaching the front row will be adequate, but the intensity of the direct sound reaching the back row will be 18 dB less (ignoring any useful first reflections), due to the inverse square law, and the higher frequencies—important for intelligibility—will be even more down, possibly by another 10 dB, because of the grazing incidence over the heads of the audience. If a loudspeaker is now placed 6 m above the head of the man speaking, the sound from it reaching the back row will be 11 dB less than the sound reaching the front row (25 m distant compared with 6·7 m), and the loss due to grazing incidence will be negligible. Thus the difference between front and back for the higher frequencies has been reduced from 28 dB to 11 dB (ignoring the contribution from the real voice). This is an extreme and simple example, but does illustrate the advantages of a single loudspeaker. Incidentally, this would be described as a 'high-level' system, which is the term for a system employing a very few loudspeakers each operated at a fairly high *level of sound* ('level' does not refer to the height of the loudspeakers above ground) and each covering a large number of people. A low-level system employs many loudspeakers each operating at a low level of sound and each covering only a few people. The comparative advantages and disadvantages of each type of system will be apparent later.

A listener in a hall with a speech-reinforcement system will receive the speech sounds at least twice, in the sense that he will receive the 'natural' speech directly from the man speaking and also the same speech coming from the loudspeakers. The loud-speaker sound will arrive at a listener before or after the 'natural' speech depending on the relative position of the man speaking and the loudspeaker, and it might be louder or softer than the natural speech, but usually louder. If there are several loudspeakers then of course a listener will receive the speech several times. The relationship in time and loudness of these repetitions (as we will call them) of the speech sounds is most important. The permissible limits are given below, but briefly if the repetitions arrive at a listener close enough together—i.e. spaced over not more than about 35 milliseconds, then they all add together and increase the loudness without decreasing the intelligibility. If on the other hand they are spaced out over a larger time interval they begin to interfere with the intelligibility; the greater the time interval between them the more they interfere, and if the time interval is long enough—about 100 milliseconds—they will be heard as discrete echoes.

There is a second effect to be considered, often called the Haas effect. (Haas was the first to put this effect, and the effect of long-delayed echoes, on a quantitative basis.) If we consider now only the original sound and one repetition of it at the same loudness and following soon after (i.e. about 25 milliseconds later), it is the sound which arrives first at the listener that determines the apparent direction of the source. An example will make this clearer. A listener is seated facing a man speaking 17 m in front of him, and has just to one side of him a loudspeaker connected to a microphone close to the man speaking. The direct sound will take 50 milliseconds to travel from the man to the listener, but the amplified sound via the microphone and loudspeaker will get to him practically instantaneously because it travels the 17 m along the cable and not through the air. If both sounds arrive at the listener at equal loudness, he will be conscious of only the first-arriving, i.e. loudspeaker sound. If we can now introduce some form of artificial delay between the microphone and the loudspeaker, for example by recording on magnetic tape the signal from the microphone and then replaying it to the loudspeaker a little later, we can arrange for the natural sound from the man speaking to arrive first at the listener. In our example, let the artificial delay be 70 milliseconds, then the direct sound will arrive first by 70 – 50 milliseconds, i.e. by 20 milliseconds. This has the very important effect of making the listener unaware of the loudspeaker, and *all* the sound—although in fact part of it is coming from the loudspeaker—appears to be coming from the man speaking. Thus the system is made to sound much more natural. This effect holds, not only when the natural and loudspeaker sounds arrive at the listener with equal loudness, but also even when the delayed sound is a good deal louder than the natural sound. In fact, the total loudness at the listener can be doubled, compared with the unaided voice, by a delayed loudspeaker without the listener being aware of the amplification (except of course that it is louder).

This technique can be extended if necessary over considerable distances. The unaided voice, let us say, will carry the first 15 m of a large hall and then needs help from the first loudspeaker. This loudspeaker will carry the voice at adequate loudness and intelligibility for another 15 m, let us say, towards the rear of the auditorium. Here, at 30 m from the man speaking, the speech requires further amplification and a second loudspeaker can now be introduced delayed so that, to a listener close to it, the first sound to arrive is that from the man speaking, the second sound to arrive is that from the first loudspeaker, and the last sound to arrive, and the loudest, is that from the second loudspeaker. Acoustically, all the speech will appear to be coming from, if not the man speaking himself, at least from the first loudspeaker, and as this will be roughly in the line between the listener and the man, it will in fact appear to be coming from the man.

A simple case involving the relative times of arrival is a rein-forcement system using one loudspeaker near to the man speak-ing. If this loudspeaker can be mounted so that it is a metre or so further from the audience than the man is, e.g. above his head, then the direct speech will arrive first and the amplified sound will not be noticeable. The trouble is that the dangers of feed-back are increased, because the loudspeaker will be nearer to the sensitive side of the microphone than if it were mounted more towards the audience and thus nearer the 'dead' side of the microphone. Nevertheless, if feed-back can be avoided it is much better to have the loudspeaker further away from the audience than the man speaking.

LOUDSPEAKERS

We should now consider how loudspeakers radiate their sound. We shall not deal with horn loudspeakers because they will seldom be used in a good-quality speech system (their main advantage being their efficiency which is not important for speech indoors) but shall confine ourselves to moving-coil loud-speakers having a cone diameter between 150 mm and 300 mm. When mounted in some conventional cabinet we can say—with-out going into too much detail—that at the lowest frequencies (up to about 200 Hz) these loudspeakers radiate equally in all directions; at the mid-frequencies (about 200 to 1000 Hz) they radiate rather less to the back but near enough equally over the front 180°; at the highest frequencies they radiate a beam of sound (which gets narrower as the frequency increases) directly towards the front, practically nothing towards the back, and little to the sides off the beam. Thus if one of these loudspeakers is used facing towards the audience, those on the axis of the loudspeakers will get all the frequencies, but those to one side will not get the higher frequencies (which are important for intelligibility). This defect is usually overcome by using several loudspeakers placed so that none of the audience is very far off the axis of one or other of the loudspeakers. An alternative method of overcoming the directional effect at the high frequencies is to use one large loudspeaker to handle the frequencies up to 500 or 1000 Hz, and another, much smaller loudspeaker, to handle the higher frequencies. The smaller the loudspeaker the less this directional effect is, so this two-speaker arrangement with its 'cross-over' connection will not become seriously directional until a much higher frequency than would the large loudspeaker on its own. It should be explained that a small (i.e. less than 150 mm diameter) loudspeaker cannot be used on its own for two reasons. The first is that it will not radiate the lowest frequencies, and the second is that it will not handle the power that will be necessary even in the smallest hall. But used in conjunction with a larger loudspeaker, it will be adequate because only some 10% of the power is contained in the frequencies above 1000 Hz.

Properly used, however, a directional loudspeaker arrangement can be a great advantage. This has been realised in recent years and has led to the development of the 'line' or 'column' loudspeaker. If several ordinary moving-coil loudspeakers are mounted one above the other (Plate X) and all are connected together in phase, by an interference effect they concentrate most of the sound into a beam which is narrow in the vertical plane but which is no more directional in the horizontal plane than an ordinary loudspeaker. This is illustrated diagrammatically in Fig. 39. An analogy is the light from a motor-car flat-beam fog-lamp.

Column loudspeakers have two big advantages over ordinary loudspeakers, and indeed over the unaided voice. The first is that they direct most of the speech sounds towards the audience. This means that the intensity of the reverberant sound is less, and thus the ratio of direct sound to reverberant sound—necessary for good intelligibility—is much improved. Secondly, the fall-off in intensity of the direct sound with distance can be reduced to some extent, by arranging the column so that the audience further away comes more into the beam.

The directivity of a column depends on its length and the frequency of the sound; the longer the column or the higher the frequency the narrower the beam. Thus we cannot get a beam of uniform width over the whole speech frequency range.

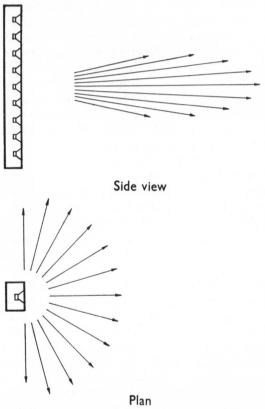

Side view

Plan

Fig. 39. Diagrammatic Illustration of Loudspeaker Column

Another effect is that when the wavelength of the sound becomes comparable with the spacing between individual loudspeakers (for speakers spaced, say 230 mm apart this happens at about 1000 Hz, i.e. when the wavelength is a little over 0·3 m) then the main beam breaks up into several beams of almost equal strength. This may not be a serious disadvantage because there will still be plenty of sound going towards the audience, and the reverberant sound is usually less at the higher frequencies. But in difficult acoustical conditions it is desirable to avoid this break-up, and to keep the directivity of the beam reasonably constant over the whole frequency range. This can best be approached by splitting the column into two parts; the long part radiating the frequencies up to, say, 1000 Hz and the short part (made up of small loudspeakers) radiating the higher frequencies. This arrangement has the added advantage of preventing the beam becoming too narrow (in the horizontal plane) at the higher frequencies, which it would do if the bigger loudspeakers were used for the whole range.

Rôle of the Operator

For any speech-reinforcement system it is advisable to have an operator skilled in the use of his particular system. In addition to the microphone switching, and any loudspeaker switching that may be necessary, his main task will be to control the volume. This is more difficult than it sounds. If only one or two people speak through the system, then the volume can probably be left at some setting found from experience, but in the more usual case of there being a constant succession of different people speaking, their deliveries will vary so much in loudness from one to the other, and from moment to moment, that continuous control is called for. The operator may have a meter to help him, which indicates the loudness of the amplified sound, but most operators prefer to rely on their ears. The greatest fault of operators is that they will have the system on too loud. There appear to be two reasons for this. The first is that the operator may feel that the system is not giving good value for money unless it is on so loud that no one can doubt that it is in use. (The ideal system, on the other hand, should be so unobtrusive that no one realises it is on, while still maintaining adequate loudness and intelligibility for everybody.) The second reason is that in any audience there are certain to be a few people who are slightly deaf (perhaps without realising it) and they might have some difficulty in hearing speech that is loud enough for everyone else. One of them may complain to the operator and ask him to turn the volume up. The operator usually complies, thus making it too loud for the rest of the audience. The only solution, and a not very satisfactory one, is to ask such people to sit closer to the loudspeakers. All operators should be warned of these two faults.

It is safer if the system is arranged so that the volume control cannot be turned up so high as to cause feed-back. This can be done by using an internal control, when the system is being finally tested, set at 6 dB below feed-back level when the main volume control on the front panel is turned full up.

Two (patented) devices are available to decrease the feed-back level by about 6 dB. The first continuously changes the phase relationship in the amplifier so that feed-back (p. 137) at one particular normal mode frequency is discouraged. The second employs sharp rejection filters at the various normal mode frequencies the system is most prone to feed-back at.

The operator and his control panel must be placed in the auditorium so that he can see the microphone positions he has to switch, and so that he hears the reinforced speech at about the average loudness for the whole auditorium. If he cannot be put in such a place, then it must be impressed upon him that most of the audience are hearing the speech so much louder or softer than he is, as the case may be.

FREQUENCY RESPONSE

All loudspeaker amplifiers are fitted with tone-controls, and it is questionable if they do more good than harm. What is certain is that in unskilled hands they can ruin an otherwise good system. It might be desirable for them to be available so that they can be set by the expert when he installs the system and thereafter not touched.

When the reverberation time of an auditorium is longer than about two seconds, e.g. in churches, cathedrals and very large auditoria, then the speech is not only more intelligible but also, surprisingly, more natural if the frequency response is limited to between about 250 Hz and 4000 Hz. This restriction must be sharp, i.e. the frequency response must fall off quickly below and above these limits. A suitable circuit for insertion in the appropriate part of the electrical system is shown in Fig. 40.

POWER REQUIREMENTS

The amounts of acoustical power required to fill auditoria with speech at normal loudness are very small. For example, in an auditorium seating about 4000, which may have a volume of about 30,000 m^3, only 0·03 watts of acoustical power is required to maintain an average level of 78 dB. This is an average figure, and the power may peak instantaneously up to ten times this value. Further, the efficiency of moving-coil loudspeakers in converting electrical power into acoustical power may not be more than 5%. So to produce the required 0·03 watts average we must multiply by ten to allow for the peaks and by, say, twenty again for the loudspeaker, and thus get 6 watts for the necessary power-handling capacity of the amplifier. This is only a small amplifier, the smallest commercial

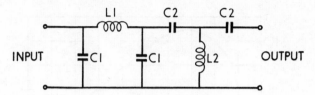

FIG. 40. Circuit to limit Frequency Response to between 250 and 4000 Hz

$$C_1 = \frac{10^3}{8\pi Z} \mu F \qquad L_1 = \frac{Z}{4\pi} mH$$

$$C_2 = \frac{2 \times 10^3}{\pi Z} \mu F \qquad L_2 = \frac{Z}{\pi} mH$$

where Z is the input and output impedance (mainly resistive)

size being usually 10 watts. A more accurate calculation of the power-handling requirements is thus not worthwhile. For music, of course, much more power is required, and also for the few situations where amplified speech has to compete with a very noisy background.

DESIGN OF TIME DELAYS

Further work has refined the original Haas results, and the effect of several repetitions of the original sounds, of modifying the frequency characteristic of the repetitions, and of the addition in loudness of several repetitions has been investigated. However, for present design purposes we need consider only two graphs. Fig. 41 shows the increase in intensity which a secondary loudspeaker (i.e. a loudspeaker repeating the speech just received from the man speaking, or a second loudspeaker repeating the speech just received from a first loudspeaker) must have compared with the primary sound if the secondary is to sound as loud. It is plotted as a function of the time interval between the arrival of the two sounds. It is seen that, for example, if the secondary sound arrives at the listener 10 milliseconds after the primary sound, then the secondary sound must be 10 dB greater intensity than the primary sound if both sound sources are to sound equally loud. In practice, of course, we do not want the two sources (which, for the front part of an audience, will be the man speaking and the nearest loudspeaker) to sound equally loud —we want the secondary sound, i.e. the loudspeaker sound, to be not noticeable. Thus the secondary sound must not be as much as 10 dB up (at this 10-millisecond delay), and under laboratory con-

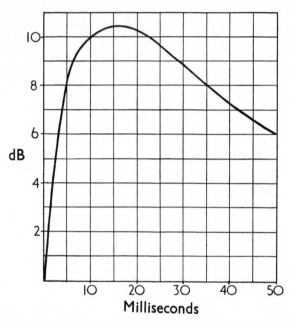

Fig. 41. Increase in Intensity (secondary sound over primary sound) for Equal Loudness as a Function of Time Interval

ditions with trained observers listening specifically to detect the secondary sound, it is known that it must not be more than 5 dB up on the primary sound if it is to be completely undetectable. However, in a practical loudspeaker system the audience is not listening so critically; further, there is the visual effect, i.e. the listeners are expecting the sound to come from the man speaking, and thus they are predisposed in favour of the sound coming from the 'correct' direction. How much can be allowed for these two factors is not known, but it is probably safe enough when designing a time-delay system to permit the secondary sounds to be 10 dB up on the primary sound over the time-interval range 5 to 25 milliseconds. Of course, if particular conditions do not call for the secondary sound to be as much as this so much the better.

The second graph (Fig. 42) necessary for the design of time-delay systems shows the percentage of listeners who will think the secondary sound is disturbing as a function of the relative intensities of the two sounds and of the time interval between them. For example, if the secondary sound is 10 dB greater than

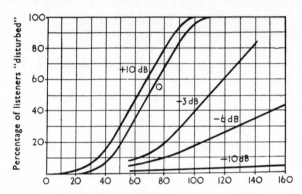

Interval (milliseconds) between primary and secondary sounds

FIG. 42. Percentage Disturbance as a Function of Relative Intensities and Time Interval

the primary sound and is 30 milliseconds behind it, then 8% of listeners will be disturbed. It should be explained that the listeners who took part in the original experiments were being very critical, and it is probably safe to say that a '10% disturbance' figure would be an adequate criterion. Thus, if the secondary sound is 10 dB greater than the primary sound then it should not be more than 30 milliseconds behind. On the other hand, if the secondary sound is, say, 6 dB *less* than the primary sound, then it can be as much as 80 milliseconds behind before 10% disturbance is reached.

The use of the two graphs can best be illustrated by a trial design. Consider the section of the hall shown in Fig. 43. Loudspeaker 1 is 3·3 m above the head level of the audience. Listener A will hear first the real voice, followed 5 milliseconds later by the sound from loudspeaker 1. We have said that the loudspeaker 1 sound can be 10 dB up on the real voice without listener A being conscious of it as a separate source. Actually, this adjustment cannot be done simply in practice, because it would involve measuring the loudness of speech sounds, so all that is done in practice is to adjust the gain of loudspeaker 1, so that the total loudness in the front area sounds correct: if the man speaks loudly, the real voice will be of the same order of loudness as loudspeaker 1, and so the real voice will predominate easily: if he speaks quietly then the loudspeaker 1 will have to be turned up as far as is necessary (feed-back permitting), and if this exceeds the permissible 10 dB difference, nothing can be done about it. But even if this state of affairs is reached, the result will still not be too disturbing, because except for those listeners very close to the man speaking, he and loudspeaker 1 will be in practically the same direction.

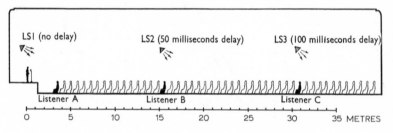

FIG. 43. Longitudinal Section of Hall with Time-Delayed Speech-Reinforcement System

As we get further away from the man speaking, the direct sound and the sound from loudspeaker 1 will fall off and further reinforcement will become necessary. Let us assume that this happens at 15 m from the man speaking; loudspeaker 2 can now be installed, let us say 3·3 m above the listeners, with an electrical delay in circuit equal to the difference in the times taken for the sounds from loudspeaker 1 and from loudspeaker 2 to reach listener B, plus an extra few milliseconds for the Haas effect. The time taken for the loudspeaker 1 sound to travel to B is about 16·5/340 seconds (340 m/sec. being the velocity of sound), i.e. 50 milliseconds, and the time taken for the loudspeaker 2 to travel is 3·3/340 seconds, i.e. 10 milliseconds. The extra milliseconds to be added depend on how many more loudspeakers are to follow No. 2. If No. 2 is the last, then the extra delay may as well be 15 milliseconds, because Fig. 41 shows that there is an optimum effect at 15 milliseconds. If there are further loudspeakers to come, then the extra delay on No. 2 should be less, because it will be necessary to fit in the later loudspeakers. Let us make the extra delay on No. 2 equal to 10 milliseconds, so the total delay introduced is $50 - 10 + 10 = 50$ milliseconds. By the time we have got as far back as this in the hall, we can assume that the real sound reaching B is negligible compared with the sound from loudspeaker 1. This is because loudspeaker 1 is operating at a much louder level than the man is speaking, and

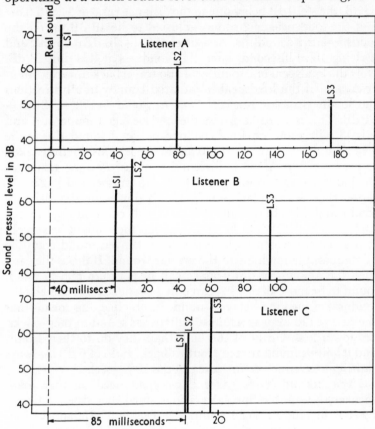

Fig. 44. Arrival of Sounds in Hall, Fig. 43

because the real sound will have been attenuated by passing at grazing incidence over the heads of the audience, compared with the sound from loudspeaker 1 which has a 'clear path' from loudspeaker 1 to B. The loudness of loudspeaker 2 can be adjusted so that its sound when it arrives at listener B is 10 dB up on the sound arriving from loudspeaker 1.

At this stage it will be useful to consider diagrammatically the arrival of the various sounds at the various positions. Fig. 44

shows that, at A, the first sound to arrive is the real sound at a level, let us assume, of 63 dB. Next arrives loudspeaker 1 sound, 5 milliseconds later, and this can be at 73 dB. The total loudness at A will be near enough 73 dB. (We will deal with the sound from loudspeaker 2 in a moment.) At B, ignoring the real sound, the sequence is : loudspeaker 1 sound (at a level of 73 − 10 dB = 63 dB, the − 10 dB being due to the distance ratio of 5·2 : 16·5 m) and the loudspeaker 2 sound adjusted to be 10 dB up on the loudspeaker 1 sound, i.e. at a level of 63 + 10 dB = 73 dB, and arriving 10 milliseconds later. The loudness at B is thus 73 dB. Now the loudspeaker 2 sound will also travel back towards listener A, and if the loudspeaker radiated equally in all directions the loudness of loudspeaker 2 at A would be 73 − 12 dB = 61 dB. This is 12 dB down on the loudspeaker 1 sound at A and it is 73 milliseconds behind the loudspeaker 1 sound, due to the electrical delay introduced (50 milliseconds) plus the time taken to get from 2 to A (38 milliseconds) minus the time taken for the loudspeaker 1 sound to get to A (15 milliseconds). Fig. 42 shows that a secondary sound 12 dB down on the primary sound and 73 milliseconds behind it will cause negligible disturbance, but it should be said here that there might be speech-reinforcement systems where this long-delayed sound could be of comparable loudness to the primary sound. If this is the case, then the amount of disturbance can be got from Fig. 42 and if found to be more than 10% can only be reduced by reducing the loudness of the long-delayed sound. In practice, this means that the delayed loudspeakers must radiate less sound to their backs, i.e. towards the front of the hall, than they do to their fronts, and it will be found that a 'front-to-back' ratio of 6 dB is always sufficient to reduce disturbance to the 10% level.

75 m beyond loudspeaker 2 we shall need another loud-speaker, No. 3. Similar calculations show that listener C will receive the loudspeaker 1 sound at a level of 57 dB, and 1 milli-second later loudspeaker 2 sound at a level of 60 dB. Thus the loudness of loudspeaker 3 arriving at listener C can be 70 dB. Loudspeaker 3 will need an electrical delay equal to that of loudspeaker 2 (50 milliseconds), plus 50 milliseconds for the distance between loudspeaker 2 and listener C, minus 10 milliseconds for the distance between loudspeaker 3 and listener C, and plus 10 milliseconds for the Haas effect, i.e. a total delay to be introduced of 100 milliseconds. The total sequence of sounds arriving at listener C is shown in Fig. 44.

It should be noted that the loudness at listener C is 70 dB, compared with that at listener A of 73 dB. We have thus got 30 m away from the man speaking for a drop in loudness of the apparent direct sound of only 3 dB, and if the system is made 1 dB too loud for A, it will be only 2 dB too quiet at C. These differences would not be noticeable, and we have still maintained the illusion that all the sound is coming from the correct direction. For comparison, if only loudspeaker 1 were used, the level at A would still be 73 dB, but at C it would be down by 16 dB.

We have been considering only listeners A, B and C in our example, but a moment's thought will show that if the system is correct for them it will be correct for the listeners behind them. This is because the relative times of arrival will remain the same, and the amplitude of the first received sound, e.g. loudspeaker 1 for listener B, will fall off more slowly as we go towards the back than will the loudness of loudspeaker 2. Thus the difference between the two loudnesses will decrease, and the Haas effect will still be all right.

If the same type of loudspeaker is used at all positions—as will often be the case—then the relative levels can be set by adjusting the voltages applied to the loudspeakers. In the above example, loudspeaker 1 had to produce a level of 73 dB at position A which was 5·2 m away. Thus at a standard distance of, say, 1·5 m, loudspeaker 1 would have to produce a level of 73 dB + 11 dB = 84 dB, the 11 dB being due to the distance ratio of 5·2:1·5. Similarly, loudspeaker 2 had to produce a level of 73 dB at B, this time at a distance of 3·3 m, so at 1·5 m loudspeaker 2 would have to produce a level of 73 + 7 = 80 dB. Thus when the system is set up originally some convenient test voltage should be inserted early in the circuit and the voltage appearing across loudspeaker 2 should be set to be 4 dB lower than the voltage across loudspeaker 1. They will of course be operating from separate power amplifiers because they have different time-delays, so this voltage adjustment is easily made. Once all the loudspeakers have been set correctly, there should be no further need to change them; the gain of the whole system should go up or down as a whole depending on the man speaking.

Time-delay mechanisms are available commercially, and Plate XI shows one make. The speech is recorded on the rotating magnetic disc, picked up by heads spaced at intervals round the circumference and then erased. The positions of the pick-up heads are easily adjustable to cover any situation met with in practice.

Design of Loudspeaker Columns

The requirements of a good loudspeaker column are (a) that it should be directional in the vertical plane, (b) that this directionality should not vary more than is possible with frequency, (c) that it should be non-directional in the horizontal plane, (d) that its radiation to the back should be at least 6 dB down on the radiation to the front, at all frequencies, and (e) that it should be as small as possible.

The directivity of a column depends on its length and on the frequency of the sound: the greater the length and the higher the frequency the greater the directivity. In detail, the directional characteristic of a source consisting of a number n, of equal point sources radiating in phase, located on a straight line and separated by equal distances d, is given by

$$R_\theta = \frac{\sin\left[(n\pi d/\lambda)\sin\theta\right]}{n\sin\left[(\pi d/\lambda)\sin\theta\right]} \quad . \quad . \quad . \quad (4)$$

where, at a large fixed distance from the source, R_θ is the ratio of the pressure at an angle θ to the pressure for an angle $\theta = 0$ (the direction $\theta = 0$ is at right angles (normal) to the line), and where λ is the wavelength. In the limiting case where n approaches infinity and d approaches zero so that $nd = l =$ the length of the line, we have the ideal straight-line source. Equation (4) then becomes

$$R_\theta = \frac{\sin\left[(\pi l/\lambda)\sin\theta\right]}{(\pi l/\lambda)\sin\theta} \quad . \quad . \quad . \quad (5)$$

In the practical cases where the source is made up of a number of loudspeakers mounted close together, formula (5) can be used with sufficient accuracy provided that the distance between the loudspeakers is small compared with the wavelength. The polar diagram (up to 30° either side of the axis) in the vertical plane of a column 3·3 m long at 1000 Hz is shown in Fig. 45. It is seen that there are secondary lobes, the greatest of them being 13 dB below the main lobe. If the line source is 'tapered' in strength so that the sound from each element varies linearly from a maximum at the centre to zero at either end, the directionality is given by

$$R_\theta = \frac{\sin^2\left[(\pi l/2\lambda)\sin\theta\right]}{\left[(\pi l/2\lambda)\sin\theta\right]^2} \quad . \quad . \quad . \quad (6)$$

Fig. 45 also shows the directionality at 1000 Hz of a 3·3 m tapered source, and it is seen that the main lobe is slightly broader while the first of the secondary lobes is reduced to 27 dB below the main lobe. The advantages of this relative suppression of the subsidiary lobes are, first, that less of the column's energy goes away from the audience and therefore there is less reverberant sound, and secondly that there may be some surface in a building, e.g. a dome, which might focus back to the audience one of the subsidiary lobes, thus perhaps causing an echo. As the

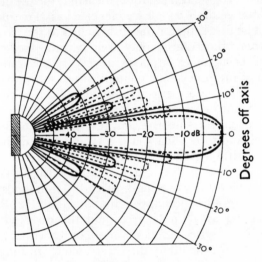

Fig. 45. Directivity of 3·3 m Column at
1000 Hz

tapering is simply done, either with a tapped transformer or with resistances, and as the power-handling requirements of any column used for speech indoors are so moderate, it seems better to use tapered columns. Other forms of tapering are of course possible, e.g. following a binomial law, which should suppress the side lobes even further, but there is so far no practical experience of them. When linear tapering is used, the voltages at the ends of the column would in theory be reduced to zero. There is obviously no point in having a loudspeaker at each end of the column with no voltage applied to it, so the tapering in practice is worked out assuming an imaginary loudspeaker at either end. For example, a column consisting of a central loudspeaker with four loudspeakers either side of it would be tapered so that the loudspeaker next to the centre one (and of course the corresponding one on the other side of the centre) would have 80% of the voltage on the centre one, the next 60%, the next 40%, and the end ones 20%.

The smaller the individual loudspeakers used in a column and thus the greater number that can be got into a given length, the nearer its behaviour approaches the 'ideal' line source. It has been found in practice that ordinary 150 mm loudspeakers used in a column are indistinguishable in quality from the best 250 mm loudspeakers, and therefore the 150 mm size may as well be used. As has been mentioned earlier, the use of the whole length of the column for all frequencies has three undesirable features. The first is that a stage is reached as the frequency increases where the column may well become too directional. There has been an example in a very reverberant building where the beam from such a column was so sharp that, at some positions, speech was perfectly intelligible when the listeners were seated but became unintelligible when the listeners stood up. This was an extreme case, but it does indicate that even in less critical auditoria such sharp focusing is undesirable. The second undesirable feature is that at a still higher frequency, when the spacing between individual loudspeakers becomes comparable with the wavelength (with 150 mm loudspeakers at 205 mm centres this will happen at about 1500 Hz) the column stops behaving as a line source, and the directivity pattern breaks up into a series of lobes of more or less equal amplitude. Thirdly, we want the minimum directionality in the horizontal plane, and even 150 mm loudspeakers will become appreciably directional in this plane at the higher frequencies.

All these defects can be much reduced by splitting the column into two parts: the whole length will be used for the lower part of the frequency range and the shorter length for the higher part. A successful arrangement is to make the whole length of the column using 150 mm loudspeakers, and to cross-over at 1000 Hz to a column of quarter the length made of 60 to 75 mm loudspeakers. Such a column, with the front grille removed, is shown in Plate X.

The simple cross-over network shown in Fig. 46 is adequate, where

$$L = \frac{R_0 10^3}{\sqrt{2}.\pi f} \text{ millihenries}$$

and

$$C = \frac{10^6}{2\sqrt{2}.\pi f R_0} \text{ microfarads}$$

where $R_0 =$ the characteristic impedance (mainly resistive), and $f =$ the cross-over frequency, usually 1000 Hz.

(Attempts have been made to reduce the length of the column with increasing frequency, thus keeping the directivity constant over the whole frequency range, by circuits which cut out the end loudspeakers successively. However, the phase changes produced by these circuits upset the directivity pattern, and it appears that this is not a practical solution.)

Another, minor, advantage of splitting the column into two parts is that as the beam gets narrower with increasing frequency so the intensity on the axis increases. The effect of this is of course halved by the splitting.

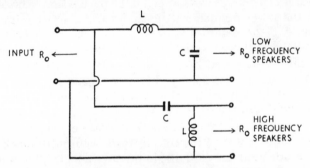

Fig. 46. Cross-over Network for Columns

All the loudspeakers must be in phase, and it is desirable to test each one, even if they all come from one batch, to ensure that their wiring is consistent. This is simply done by applying a small d.c. voltage to the loudspeaker and seeing or feeling the direction of motion of the coil. Also, all the loudspeakers should have equal acoustical efficiency, and while speakers of the same type do not usually vary much, for complete safety—and to ensure against any defective ones—a simple comparative listening test with a steady signal applied (speech or dance music) is adequate.

The two parts of a column will probably have different acoustical efficiencies because of the different sizes of loudspeakers used. Usually the larger loudspeakers will be more efficient, and thus either the transformer tappings should be adjusted to compensate for this, or an attenuator should be fitted in the low-frequency part.

Other points about design are: (a) separate pairs should be run to each loudspeaker in the column; otherwise (i.e. if a common return is used) the voltage drop along this common return may be sufficient (due to the comparatively heavy current taken by the centre loudspeakers) to upset the voltages fed to the loudspeakers near the ends of the column; (b) the smaller loudspeakers must be mounted flush with the surface, otherwise their radiation to the sides may be restricted; (c) a resonance may occur behind the loudspeakers between the two parallel sides of the column, and this can be stopped by making these sides non-parallel internally, i.e. by fitting an oblique wood block along the length of one side.

It is nearly always desirable for the column to radiate less sound towards its back than it does towards its front. At the higher frequencies this will happen if the back is simply boxed in, but at the lower frequencies the sound will diffract round the column. A better design is to use for the back a layer of cotton-waste. Briefly, the effect is that the back radiation from the loudspeakers is delayed by the cotton-waste by a time equal to the time taken for the front radiation to reach the back round the sides of the column. As the back radiation is of course 180° out of phase with the front radiation, the total radiation to the back is reduced by cancellation. With careful selection of the thickness and density of the cotton-waste the front-to-back ratio can be made as much as 20 dB; in practice a 25 to 50 mm thickness of cotton-waste will give a ratio of at least 6 dB at all frequencies, which is sufficient in most cases.

Use of Loudspeaker Columns

Columns are made commercially in lengths ranging from 1·2 m to 3·3 m, and while many of the commercial types use only one line of loudspeakers for the whole frequency range—with the disadvantages mentioned earlier—there are makes available which have the cross-over arrangement. How long a column is necessary for a given hall cannot be stated with any certainty: it is largely a matter of judgement. A very rough guide would be:

Distance to be Covered by Column (m)	Reverberation Time at 500 Hz (hall full) (secs)	Length of Column Required (m)
30	4 or longer	3·3
15	4 or longer	3·3 or 2·4
30	2 to 4	2·4
15	2 to 4	2·4 or 1·8
30	1·5 to 2	1·8
15	1·5 to 2	1·2 or 1·8

The two acoustical factors governing the orientation of loud-speaker columns are, first, that the main beam of sound should fall on the audience area, not only for the obvious reason that this is where the sound is required, but also so that this main sound is absorbed by the audience and does not become un-wanted reverberant sound, and secondly, that the directivity of the column should be used to compensate as far as possible for the fall-off in intensity with distance. To some extent these two requirements work against each other, as will be shown.

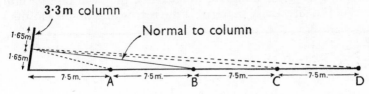

FIG. 47. Effect of Distance and Directivity

Consider, for example, a hall with a flat floor 30 m long and employing a 3·3 m linearly-tapered column. This is illustrated in Fig. 47. The column could be placed with its bottom at ear height and inclined forward so that the normal from the centre of the column meets the audience at 15 m away, i.e. half-way back. With this arrangement the main beam is falling on the audience area, and we should now consider how much distance compensation will occur. Of course, at the lower frequencies we shall get little compensation because the beam is too wide. But it is the higher frequencies (i.e. above 1000 Hz) which are the most important for intelligibility, because the energy of con-sonants is in this range, and it is here where we can get some useful distance compensation.

For comparison we will consider the two frequencies of 250 Hz and 4000 Hz. If we take listener position B (Fig. 47) as our reference point, we can prepare a table of the variation in loudness due to distance and directivity as follows:

| Position | 250 Hz | | | 4000 Hz | | |
	Distance Factor	Directivity Factor	Total (dB)	Distance Factor	Directivity Factor	Total (dB)
A	+6	−0·5	+5·5	+6	−6	0
B	0	0	0	0	0	0
C	−3·5	0	−3·5	−3·5	−0·5	−4
D	−6	0	−6	−6	−1·5	−7·5

Thus the range of intensities from A to D at 250 Hz is 11·5 dB, i.e. we have gained only 0·5 dB compared with the range of 12 dB a non-directional loudspeaker would give, but at 4000 Hz the range is 7·5 dB, i.e. an improvement of 4·5 dB.

If the column is pointed still further back, then an appreciable amount of the sound energy in the main beam will miss the audience area and will strike the rear wall, after which, if the rear wall is reflecting, it will become unwanted reverberant sound. Further, the front row of the listeners might be so far off the axis that, for the highest frequencies, they will be in a null area of the directivity pattern. If the rear wall is absorbent then

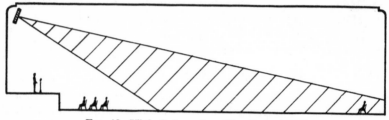

Fig. 48. High Column in a Flat-floored Hall

the first point is not important; if the front listeners get enough sound from the man speaking then the second point is not important. In general, the best compromise is to point the loudspeaker column about two-thirds to three-quarters of the way back.

In some cases it might be more convenient to sacrifice the compensation for distance, and mount the column fairly high up (as illustrated in Fig. 48) and pointed so that all the main beam strikes the audience. The main advantage of a column—its reduction of reverberant sound—is thus kept, and feed-back troubles may be made easier because the microphone is well out of the column's main beam.

If some part of the building, e.g. a pillar, comes between the loudspeaker and the listener then it will throw an acoustic 'shadow'. How serious this is depends on the size of the obstruction, but every effort should be made to avoid having listeners behind any object more than 300 mm wide.

THE USE OF SPEECH-REINFORCEMENT SYSTEMS IN SOME TYPICAL AUDITORIA

We shall here consider the designs of speech-reinforcement systems for a few typical auditoria, in the light of the above discussions. It is difficult to say exactly what size auditorium will need a speech-reinforcement system. A hall used for a variety of purposes, e.g. a school hall, with a flat floor, with some intruding noise and often with indifferent speakers, will probably need one if it holds 500 or more audience. On the other hand, a theatre with a raked floor, with quiet conditions and with trained actors (who manage to speak about 10 dB louder than untrained people without obvious effort), could probably hold up to 1500 audience before a reinforcement system was necessary. There is no doubt that in a new building provision should be made for a speech-reinforcement system if there is any possibility of one being required. Otherwise, there is a danger of temporary systems with bad loudspeaker placing being introduced at a later date.

We should distinguish between two classes of auditoria. The first is where the acoustics are good, i.e. the reverberation time is correct for speech and there are no specific acoustic defects such as echoes or 'dead' areas; the second is where the acoustics are bad. For the good auditoria, a speech-reinforcement system will only be needed to make the speech louder, particularly at the back of the auditorium. An example is a hall with a flat floor seating 500 audience with a man speaking from the stage. His unaided voice will not reach to the back because of the attenuation due to distance and grazing incidence. A single loudspeaker placed a few metres above the head of the man speaking will make things much better because the grazing incidence will be removed and because it can be operated at a higher level than the man can speak without being too loud for the front rows.

At the higher frequencies a single loudspeaker may be so directional as to reduce the intelligibility for those off the axis of the loudspeaker, i.e. in this case those seated near the front and to either side. Further, feed-back may be troublesome with the loudspeaker only a metre above the microphone, although this can be minimised (if a ribbon microphone is used) by mounting the microphone horizontally. The most common method of overcoming these two disadvantages is to have two loudspeakers on either side of the centre-line, i.e. either side of the proscenium arch. Thus nobody will be very far off the axis of one or other of the loudspeakers. The loudspeakers will be further from the microphone and feed-back will be reduced, but there are two disadvantages. The first is that those seated on the centre-line of the hall will receive two loudspeaker sounds. The real sound will arrive first, and if it is loud enough it will determine the apparent direction of the speech as coming from the man speaking, which is good. If, however, it is not loud enough, the speech will appear to be coming from either one side or the other, depending which way the listener leans in his seat. This can be irritating. If there is a centre gangway then the problem does not arise. The second disadvantage is that for those of the audience who are nearer to either of the loudspeakers than they are to the man speaking the speech will appear to be coming from the loudspeaker. On the whole, if feed-back can be avoided, probably the best solution is to have one central loud-

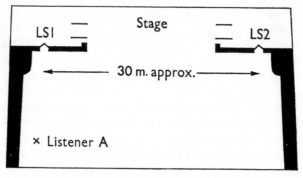

FIG. 49. Plan of Theatre with Two Loudspeakers

speaker unit with a cross-over network and a small loudspeaker to overcome the directivity.

If two loudspeakers either side of a stage are used and if the auditorium is large enough and is fan-shaped or horseshoe-shaped, then the time interval between the arrival of the two lots of loudspeaker sound might be long enough to cause some interference with the speech. For example, consider the plan shown in Fig. 49. The sound from loudspeaker 2 will arrive at listener A about 75 milliseconds after the sound from loudspeaker 1, and as there is a factor of two in the distances listener A is from the two loudspeakers, the sound from loudspeaker 2 will be 6 dB down on the loudspeaker 1 sound. Fig. 42 shows that when the secondary sound is 6 dB down on the primary sound and is 75 milliseconds later, then the disturbance figure is about 10%. Only under severe conditions, e.g. rapid speech in a rather reverberant auditorium, would this disturbance figure be significant, and this example has only been given because such conditions did once occur in the authors' experience, giving rise to severe complaints from some members of the audience. The trouble was stopped by replacing the two loudspeakers by one over the centre of the proscenium arch.

Our second example is a hall with otherwise good acoustics, but with a bad area, such as often occurs underneath a deep balcony. This can be dealt with by placing loudspeakers in the soffit of the balcony, but it is then essential for realism and often for intelligibility that these loudspeakers should be time-delayed. Several examples have occurred where such loudspeakers without time-delays have been installed, but were found to do more harm than good. The time-delay introduced should of course correspond to the time taken for the sound from the stage to reach this area, less the time taken for the sound from the subsidiary speakers to reach this area, *plus* about 15 milliseconds, and the amplitude of these subsidiary loudspeakers should be such that the sounds from them reaching the listeners should not be more than 10 dB up on the sounds reaching them from the front of the hall.

We have so far been considering only halls whose acoustics are good. When we come to deal with halls with bad acoustics (and this usually means their reverberation time is too long), then there are two solutions. The first is a low-level system, i.e. a large number of small loudspeakers distributed so that every person in the audience is close enough to a loudspeaker for the direct sound—from this nearest loudspeaker—to predominate over the reverberant sound. This is the most fool-proof system; in the extreme case—as for example is done in the 6000 seater Palace of the Soviets in Moscow—each member of the audience has his own loudspeaker and is thus so close to it that the room acoustics do not matter. However, this type of system hardly comes within our definition of high quality; although the intelligibility will be excellent—if there are enough loudspeakers—there will be no realism because the speech will be coming from the nearest loudspeaker and, further, the large number of loudspeakers all operating at the same level produce a rather unpleasant-sounding effect. To maintain realism—where this is important—it is better to use loudspeaker columns of the appropriate length (p. 158) in difficult acoustical conditions.

On p. 159 the use of loudspeaker columns to cover a flat area was described. Similar reasonings apply if the auditorium floor is raked, and if there is a balcony, two columns may be used, one to cover the ground area and the other to cover the balcony, as illustrated diagrammatically in Fig. 50. The two columns should be a few metres apart to minimise interaction between them, but should not be so far apart (not, say, more than 10 m) as to cause trouble due to the time interval between arrival of the two sounds in areas of the auditorium covered by both columns.

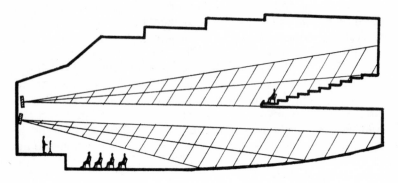

FIG. 50. Use of Two Columns in a Hall with Balcony

Probably the worst buildings for speech are large churches and cathedrals. Their reverberation times are long, and the distances to be covered by the loudspeakers are great. Further, the system has to work not only when there is a full congregation but also when the floor area is only sparsely covered. Loud-speaker columns are invaluable here, but in the largest cathedrals one column will not cover the whole area, and subsidiary, time-delayed columns will also be needed. We will describe briefly the speech-reinforcement system installed in Salisbury Cathedral as our next example.

Fig. 51 shows the plan of the Cathedral. It was necessary to reinforce speech from the pulpit, the lectern, the choir and the altar. The reverberation time at 500 Hz in the full cathedral is 4 seconds, and 6·5 seconds when empty. With this long rever-beration 3·3 m columns were necessary, and one was placed close to the pulpit and one close to the lectern (to be used alternately as required). The base of each column was about 0·2 m above the ear height of the congregation and they were tilted forwards so that the axis of the beam reached the congregation about 12 m back. The acoustic 'shadows' cast by the nave pillars would affect a large proportion of the congregation, so three pairs of subsidiary columns were mounted on the nave pillars

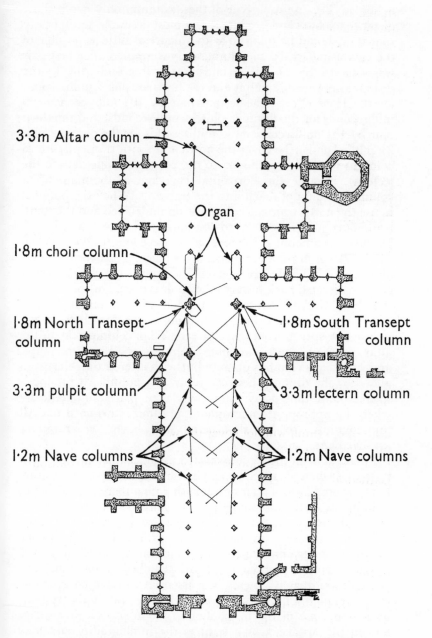

3·3m Altar column

Organ

1·8m choir column

1·8m North Transept column

3·3m pulpit column

1·8m South Transept column

3·3m lectern column

1·2m Nave columns

1·2m Nave columns

FIG. 51. Plan of Salisbury Cathedral showing Position of Loudspeakers

as shown. Because each pair of these columns only had a short distance to cover and because it was most desirable aesthetically that they should be small so as to obscure as little as possible of the famous pillars, 1·2 m columns only were used. The first pair was delayed by 57 milliseconds (50 milliseconds due to the distance between the pulpit and the first pair plus 7 milliseconds for the Haas effect), the second pair by 104 milliseconds (90 milliseconds for the distance plus 14 milliseconds) and the third pair by 151 milliseconds (130 milliseconds for the distance plus 21 milliseconds). To cover the north and south transepts, 1·8 m columns were used (one next to the pulpit and one next to the lectern), because the horizontal coverage of the main 3·3 m columns would not reach into the transepts. When the pulpit is in use the north transept column is on undelayed and the south transept column is on delayed—as happened to be convenient— by the same amount as the first pair of nave loudspeakers. When the lectern is in use the reverse takes place. To cover the choir area (there is a glass screen across the front of the choir) a 1·8 m column directed back towards the choir is used. For speech from the choir, this column is disconnected and the pulpit column and the subsidiary columns function as for speech from the pulpit. For speech from the altar, the 3·3 m column near to the altar comes on, the 1·8 m choir column goes off, and the pulpit and subsidiary columns operate but with altered time-delays to allow for the distance between altar and pulpit. All the loud-speaker and time-delay alterations take place automatically when the appropriate microphone position is switched on. All the columns are linearly-tapered and have the cross-over arrangement.

Finally, one of the most awkward of all speech-reinforcement systems is when the speech may be from more or less any position in the room. This happens in debating chambers, sometimes in banqueting halls and sometimes in halls used for political or other discussions when the audience participates. The usual solution is to use a low-level system with several microphone positions, or a trailing microphone, and usually the loudspeakers closest to the microphone in use are switched off or reduced in volume. With such systems it is essential to have an operator in attendance all the time, versed in the use of the system; otherwise some of the microphones not in use are invariably left on by mistake, which reduces the intelligibility and may cause feed-back. It should be possible in theory to divide such a room into several areas, each with its own microphone and loudspeaker, and arrange an automatic time-delay and amplitude-adjusting mechanism so that, depending on which microphone is in use, the time and amplitude sequence is nearly correct for any part of the room. A system based on this principle is operating successfully in the South African Houses of Parliament.

7

General Principles of Sound Insulation and Noise Control

All noise-control problems involve three parts: the noise source, the recipient of the noise, and the 'path' between the two. This 'path' may be simple, such as the air between a machine and a man standing close to it; or it may be complicated, such as the varying air, the roof, walls and windows, and the air again between an aircraft in flight and a man in a room. Noise sources may be indoors or out-of-doors, but as far as this book is concerned the recipients of the noise are always indoors.

It is always necessary to consider noise problems in terms of frequency, because sources, 'paths' and effects on recipients are all frequency-dependent.

A noise source may be either non-directional, i.e. it may radiate noise equally in all directions, or it may be directional, i.e. it may radiate more noise in some directions than it does in others. (And, as we have said, all characteristics—such as directionality—will vary with frequency.) For noise sources operating in rooms (using 'room' in its technical sense of an enclosure) it is seldom that any directionality a source may have is very important; on the other hand, the directionality of noise source in the open air is often a most important factor.

Noises in the Open Air

The possible defences against noise sources in the open air are: to put as great a distance as possible between the source and the building to be protected; to put a screen, e.g. a wall, between the source and the building; or to design or position the building itself to 'resist' the noise, e.g. double windows. (Of course, it is sometimes possible to box in the source completely, but then we would no longer call this a noise source out-of-doors.)

Any directivity a noise source may have is important. For example, a jet engine radiates most of its low-frequency noise at an angle of about 30° to the jet axis. Thus, if the noise from a jet engine is to be measured, the measurements must be made at several positions round the engine if a complete picture is to be obtained. Fig. 52 illustrates this point and shows the sound-pressure levels in the 300–600 Hz octave band at various

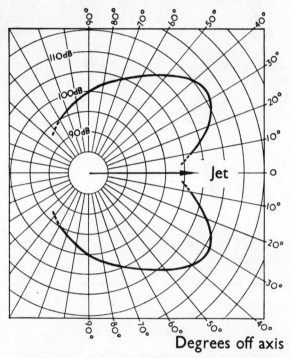

Fig. 52. Noise Levels in 300–600 Hz Octave of a Jet
Engine on the Ground at a Distance of 30 m

positions round a jet engine running in the open air. Obviously a building situated at about 30° to the jet axis is going to receive more noise than if it were situated at, say, 90° to the axis.

The first factor determining the reduction of intensity of a noise with distance is the inverse square law, i.e. a drop of 6 dB every time the distance between the source and the receiving point is doubled. The second is the molecular absorption of sound by air. The detailed values of this absorption are given in Chapter 9, but it should be noted that at frequencies below 1000 Hz this attenuation is negligible. It is often the case that the molecular absorption predominates over all other attenuating factors, and this means that at fair distances from the source we are left with only the frequencies below 1000 Hz to consider.

Apart from this molecular absorption, the other weather factors that affect sound propagation are wind and temperature *gradients*, snow and fog. It is a common misconception that sound is attenuated merely because it travels up-wind. The true action is that when there is a wind there is always a wind gradient. This is illustrated in Fig. 53. The wind at the bottom is not blowing as fast as it is at the top, and this means that the sound waves travelling up-wind near the bottom are moving faster than those at the top. The sound waves thus bend round as shown, resulting in less energy near B than there would be otherwise.

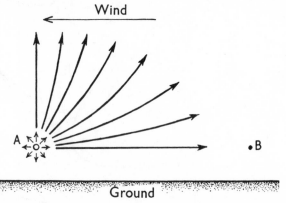

Fig. 53. Illustration of Effect of Wind Gradient on the Propagation of Sound

Temperature gradients have a similar effect. This is because the velocity of sound increases with increase in temperature. Thus, if the temperature of the air is higher near the ground than it is in the upper layers (the usual case during the day) the sound waves higher above the ground will travel slower and the sound rays will be bent upwards. Thus, all the way round the source the sound will be quieter than it would have been with no temperature gradient. Conversely, when the temperature is lower near the ground than it is in the upper layers (the usual case during the night) the sound waves higher above the ground will travel faster and the sound rays will be bent towards the ground. Thus all the way round the source the sound will be louder than it would have been with no temperature gradient.

Note that the temperature gradient affects the sound in all directions round the source, unlike the wind gradient which has different effects down-wind and up-wind (and has no effect at right angles to the wind direction).

It is usually wind and temperature gradients effects that account for the occasional freak reception of sounds over very long distances (usually from explosions) while places nearer to the source of sound hear nothing. The noise has been bent upwards by a gradient and after travelling long distances in the upper atmosphere has been bent down again by a reverse gradient.

The effects of fog and snow are mentioned here merely for the sake of completeness: they will not normally be factors of importance in any design problem. Fog causes an extra absorption

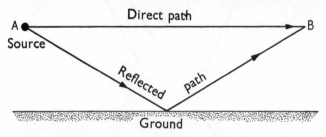

FIG. 54. Illustration of Reflected Ground Wave

to take place in the air and for a moderately dense fog (visibility 46 m) this extra attenuation is of the order of 10 to 33 dB/1000 m, depending on the frequency. Snow forms an absorbent layer on the ground and this affects the reflected ground wave (see below), the general effect being to reduce the sound level at a distance.

When the source of sound and the receiving point are not far above the ground then the reflected wave off the ground becomes important. This is illustrated in Fig. 54. It is obvious that the nature of the ground, e.g. hard and reflecting or soft and absorbent, will affect the reflected sound and thus the total sound received at B. The exact effect of this ground reflection is most complex and will not be dealt with here, except to mention that a hard ground surface such as concrete will not cause any extra attenuation but a soft surface such as grass will.

Belts of trees will cause some attenuation of sounds, but they often also affect the wind gradients and this may counteract the additional attenuation.

Turbulence in the atmosphere, i.e. eddies of air of various sizes from centimetres up to a few metres in diameter which occur to a greater or lesser extent on most days except the stillest, do not have much effect on the propagation of sound. Probably because of local varying wind and temperature gradients they do cause the sound level at the receiving point to fluctuate up and down, but the average effect is little. These second-to-second fluctuations are often noticeable when listening, for example, to aircraft noise over some distance. When the noise source is near the ground, then all the factors described above are involved. However, when the noise is well above the ground —which in practice means aircraft—then conditions are rather simplified. When the sound is propagated vertically, i.e. from

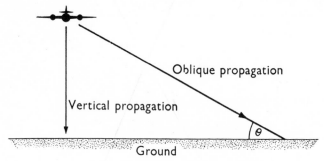

Fig. 55. Oblique Propagation of Sound

an aircraft in flight to a point immediately underneath, the propagation cannot be affected by wind and temperature because the wind or temperature gradients are in the same direction as the sound propagation and, as has been shown, it is only such gradients at right-angles to the direction of propagation of the sound that can affect it. The only meteorological factor of importance is then molecular absorption. For oblique propagation (i.e. see Fig. 55) it is also probable that the only important factor is the molecular absorption until the aircraft is so low or the horizontal distance is so large as to make the angle θ on Fig. 55 as small as 5° to 10°.

It depends on how detailed a solution to a particular problem is required whether or not wind and temperature effects need be considered. For a long-term average the wind effect can be ignored. Similarly, temperature gradients will change usually between day and night, and if it is possible to average day and night conditions then the average temperature effect will also be negligible. Of course, it is often not permissible to lump day and night conditions together like this.

Naturally, any obstruction such as a wall will reduce the propagation of sound. However, in general the obstruction must be large compared with the wavelength of the sound if it is to have much effect. For example, if a noise of 100 Hz is involved the wavelength is 3·3 m and a wall built to obstruct this noise by about 20 dB would need to be of the order of 7·6 m high, if the source is 3 m from the wall. The attenuation is so limited because the sound diffracts over and round the obstacle, because

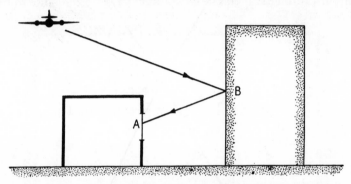

FIG. 56. Reflection of Sound off Outside Wall

it is scattered by the air above the wall and because the ground attenuation is reduced (see p. 252). The wall itself need only be of quite flimsy construction, because there is no point in its sound insulation being any more than the limited attenuation. Thus this particular wall need weigh no more than 49 Kg/m² (it must of course have no holes in it).

Surfaces in the open air will reflect or absorb sound in the same way as they do indoors. The number of types of surface suitable for outdoor use is very small and they are nearly all hard and solid and therefore sound-reflecting. It is sometimes useful to have an outside surface sound-absorbing; an example is shown in Fig. 56, where the noise reaching window A from the aircraft would be much less if the facing surface B were absorbent. It is difficult to make a weatherproof absorbent surface, but some such construction as steel wool behind a perforated metal facing—all duly treated—might last for a few years, and various sprayed materials have been used successfully.

NOISE SOURCES INDOORS

When a noise source is operating in a room, the noise level at any position may be regarded as being made up of two parts; the first part is the direct sound, i.e. the sound travelling directly from the source to the position under consideration, and the second part is the reverberant sound, i.e. the sound reaching the position after multiple reflection from the room surfaces. This is illustrated in Fig. 57. The intensity of the direct sound will fall off as the distance from the source increases, by 6 dB every time the distance is doubled. On the other hand, the intensity of the reverberant sound will, near enough for present purposes, be uniform throughout the room volume. Close to the noise source

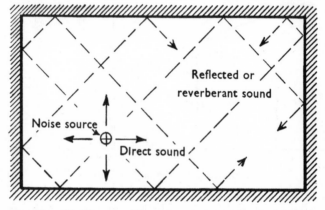

FIG. 57. Direct and Reverberant Sound in a Room

the direct sound will predominate but remote from the source the reverberant sound will predominate. The intensity of the sound will thus change with distance from the source as illustrated in Fig. 58. It follows that when measurements of noise level are made in a room it may be necessary to determine whether it is predominantly the direct or the reverberant sound which is being measured.

It depends on what the problem is whether we are more interested in the direct sound or in the reverberant sound. If it is a question of the noise being transmitted from one room to another then it will nearly always be the reverberant sound which is of interest, because it is this which is incident on the room surfaces and so transmitted to other rooms. If it is a question of protecting people in the same room as the noise then either the direct or the reverberant sound may be important.

The relative levels of direct and reverberant sound can be calculated if sufficient is known about the noise source and the room acoustics. This is seldom the case and as a working rule, but with some exceptions, it can be said that the direct sound predominates for a metre or so from the source. In practice this means that anyone operating a machine (whether it be a machine in a factory such as a pneumatic drill or a machine in an office such as a typewriter) will be in the area where the direct sound predominates.

It is obvious that nothing done to the room itself will affect this direct sound. If, for example, in a factory the noise is so loud as to cause permanent deafness then, assuming the machine itself cannot be modified so as to produce less noise, the operator

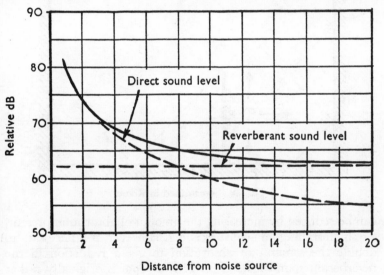

Fig. 58. Decrease in Intensity of Sound with Distance from Source

of the machine should wear ear-plugs. If there is no danger of deafness it is doubtful if it is necessary to worry about the direct sound. Most people have no objection to the noise that they themselves are making, and in many industrial operations it is a positive advantage for the operator to hear his own machine clearly.

However, there are some cases where screens are useful, particularly if the noise source is directional. Consider the room shown in Fig. 59. The screen indicated is absorbent on the side facing the noise source. The direct sound falling on this face will be mainly absorbed and will thus be stopped from contributing to the reverberant sound. Further, any person behind the screen will be protected from the direct sound (provided the dimensions of the screen are large compared with the wavelength of the sound).

Turning now to the reverberant sound, its intensity for a given noise source will be determined solely by the amount of absorption present in the room (not by the room volume, although in practice the amount of absorption usually increases with increasing room size). Thus the intensity of the reverberant sound

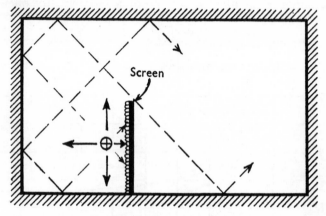

Fig. 59. Screen used in Room

can be reduced by increasing the amount of absorption present, e.g. by installing an acoustic ceiling. However, it is necessary to double the amount of absorption to get a reduction in the reverberant sound level of 3 dB. If a room is originally rather bare the installation of an acoustic ceiling may increase the absorption present by a factor of, say, ten, which will reduce the reverberant sound by 10 dB, but to get a further 3 dB reduction would need further absorption equal to the absorption of the ceiling plus the absorption originally present, which is usually impossible to achieve.

The advantages of introducing as much absorbent as is convenient into a room and thereby reducing the intensity of the reverberant sound are, first, that the room is quieter except for those people in the direct sound field. Secondly, the sound energy falling on the room surfaces is less and this means less sound will be transmitted to other rooms. Thirdly, an operator will probably hear his own machine clearer if the noise from other machines in the same room is less. This is particularly important if the machines are different and some are noisier than others; the operator of a quiet machine may not be so disturbed by the noise from a remoter but noisier machine. Fourthly, if the noises are transient, e.g. a burst of riveting in a workshop, then the succeeding reverberation is reduced. Finally, the impression of general clangour is lessened, and some people are sensitive to this, although there is little evidence that their efficiency is affected one way or the other.

In long low rooms with an absorbent ceiling the reverberant sound is not uniform and noise generated at one end of the room will be reduced in intensity the further one gets away from the source. This may have a practical application in, for example, a large general workshop where very noisy operations such as riveting can be kept to one end.

SOUND INSULATION

We now come to the problems of noise getting from one room to another, i.e. the problems of sound insulation. We have already said that extra absorption in the room where the noise is will reduce the intensity of the reverberant sound and thus help the insulation, but we should stress at once that the reduction of intensity got in this way is usually small compared with the reduction got by sound insulation. In other words, extra sound *absorption* is no substitute for adequate sound *insulation*. For example, if the room is originally bare, as a workshop might be, an acoustic ceiling might reduce the intensity in this source room by as much as 10 dB, but this should be compared with the 50 dB sound insulation of a 220 mm brick wall. This 10 dB extra reduction is, however, well worth while in conjunction with an insulating division, because to raise the insulation of a wall by 10 dB it would be necessary to quadruple its weight. On the other hand, if the room has originally a fair amount of absorption in it, as for example has a furnished room in a flat, then an acoustic ceiling would, at the most, reduce the intensity by 3 dB.

This point is stressed here because it is such a common misconception that sound absorption will solve all noise problems.

The room in which the noise originates will be called the source room, and the room to which the noise is being transmitted will be called the receiving room.

There are two types of sound insulation to be considered: air-borne sound insulation and impact sound insulation. The first type concerns insulation against noises originating in the air, e.g. voices; the second type concerns impact noises, e.g. footsteps. This second type of noise is really a combination of airborne and impact noise, because the impacts will produce air-borne noise in the source-room and this air-borne noise will also be transmitted. But in nearly all cases the noise produced in the receiving room by the direct impact noise predominates.

Further, vibration generated in one room may set the surfaces of a remote room into vibration and they will thus radiate some noise. This is discussed further below.

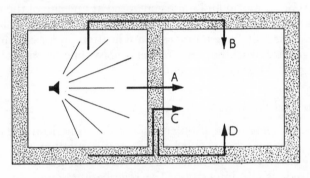

FIG. 60. Paths for Sound Transmission between Adjacent
Rooms

We will deal first with air-borne sound insulation. As has been said, a noise source operating in one room will produce a reverberant sound field which impinges on all the room surfaces, setting them into vibration. (If the noise source is very close to one of the surfaces it may be the direct sound which predominates at this surface, but this is seldom the case.) The simplest case is when the receiving room is separated from the source room by a single, solid wall with no openings of any kind. The first and most obvious way for the sound to be transmitted from one room to the other is shown as 'Path A' on Fig. 60. The sound waves falling on the source room side of this wall move it in and out (of course by very small amounts, impossible to see or feel) and it thus radiates sound on its far side, into the receiving room. The amount of radiation and hence the sound insulation will depend on the frequency of the sound and on the construction of the wall, above all on its weight. Practically all walls and floors have less insulation at the low frequencies than at the high frequencies.

For the single-leaf solid walls we are now considering the sound insulation increases by about 5 dB for every doubling of the frequency, i.e. every octave. For example: a wall whose insulation is 30 dB at 100 Hz will have an insulation of 35 dB at 200 Hz, 40 dB at 400 Hz, and so on. Thus a complete description of the insulation of a wall must give the insulation at all frequencies. For convenience, however, the average insulation over some frequency range is often used. In some circumstances this can be misleading but for present purposes we will use this average figure. The frequency range for the average is from 100 to 3150 Hz, this range covering most of the frequencies that are of practical importance. The example we have just given of a wall whose insulation is 30 dB at 100 Hz, 35 dB at 200 Hz, and so on, will have an insulation (assuming that the increase in insulation with frequency is exactly 5 dB per octave) of 55 dB at 3150 Hz. The average insulation over this frequency range will therefore be 42·5 dB. One of the dangers of this 'single figure' value for insulation is that the single figure may be different for the same wall if another frequency range is used. If, for example, the range 200 to 3150 Hz is used the single figure for this wall would be 45 dB, an apparent improvement of 2·5 dB. Until recent years the frequency range used in the United Kingdom was 200 to 2000 Hz. The single-figure averages for this range do not, however, differ by more than 1 or 2 dB from the single figure for the 100 to 3150 Hz range, because the old range is less than the new range at both the high- and the low-frequency ends, leaving the average not much affected. It is obvious that when a single figure is quoted the frequency range it refers to should always be given.

For single walls the average insulation is almost entirely determined by its weight per unit area. This is the so-called 'mass law', and it is rare to find a significant (i.e. more than ± 2 dB) departure from it. Fig. 61 shows this mass law plotted over the range of superficial weights met with in practice. It will be seen, for example, that a wall which weighs 245 Kg/m², e.g. a 110 mm brick wall, has an average insulation of 45 dB, while a wall weighing 490 Kg/m², e.g. a 220 mm brick wall, has an insulation of 50 dB. The line in Fig. 61 is not quite straight, but over the heavier part of the range it can be seen that doubling the weight gives an increase in average insulation of about 5 dB. This 5 dB increase in the average insulation per doubling the weight should not be confused with the 5 dB per octave increase described earlier.

We have so far been considering only Path A in Fig. 60. With fairly light-weight partitions, i.e. those with an insulation of 40 to 35 dB or less, this is usually the only path that matters (still assuming that there are no openings between the rooms) but at higher insulations other paths for the sound become increasingly important. These other paths are: Path B, which is due to sound falling on the surfaces of the source room other than the dividing wall, the resultant vibrations travelling along in the walls or floors into the receiving room and being radiated there; Path C, which is due to the vibrations of these other surfaces in the source room causing the dividing wall to vibrate (because they

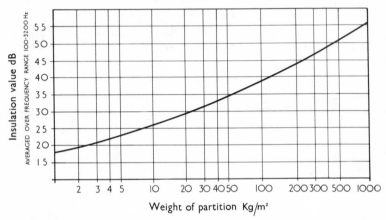

FIG. 61. 'Mass Law' of Sound Insulation

are joined to it) in addition to the Path A vibration of the dividing wall; and Path D, which is due to the vibrations of the dividing wall causing the other surfaces in the receiving room to vibrate. All these three paths, B, C and D, are lumped together under the heading of 'indirect' or 'flanking' transmission.

How important this indirect transmission is depends on the construction of the dividing wall and of the flanking walls and floors—each reacts on the others. It is doubtful if anything practicable can be done, in ordinary buildings, to reduce flanking transmission. Any flexible layers that might be used for this purpose would have to be quite thick—at least 75 mm—for them to have any effect at all. Probably the most important point about indirect transmission is that changes in the section of the wall or floor will reflect much of the sound energy in the wall or floor. For example, Fig. 62 shows the cross-section where a dividing wall meets the supporting (solid) floor. Indirect transmission by Path B will be less by 9 dB if the dividing wall is fixed solidly to the floor than it would be if the wall were not in contact with the floor. If the wall is bigger than the floor then the reduction in transmission will be more than 9 dB; if it is smaller, then the reduction will be less. On the other hand, Paths C and D will be more than they would be if some effective resilient layer were put between the wall and the floor. Further, Path A might be affected by the presence of any resilient layer.

There is not much evidence at the present time on the relative magnitude of the various paths for indirect transmission, but it

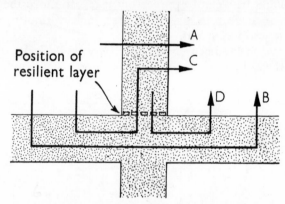

FIG. 62. Reflection of Sound Waves due to Change in Section

does appear that on the whole it is better to bond all walls and floors solidly together as a means of reducing indirect transmission. There are exceptions—some of the multiple-leaf constructions depend on a resilient layer round the edges of one of the leaves for their effectiveness—but except for these very special designs solid bonding is safer.

Although the 'mass law' provides a good working rule for the average figure for the air-borne sound insulation, the true behaviour of a partition—in particular the variation of its insulation with frequency—depends also on other factors than its weight. We will deal briefly here with one only—the coincidence effect. This is that when the projected length of the sound waves in air is the same as the length of the flexural waves in a partition then the insulation is much reduced. This happens at some frequency and at some angle of incidence with nearly all partitions: the aim should be to keep this coincidence frequency at normal incidence outside the important subjective range.

The velocity of flexural waves in a partition and hence—for a particular frequency—the wavelength varies with the stiffness of the partition. The greater the stiffness the higher is the velocity and the longer the wavelength. Thus for heavy, stiff partitions the coincidence frequency at normal incidence will be low; for example, for a 220 mm brick wall it is about 80 Hz which is usually not important subjectively. For lighter partitions the coincidence effect may come at, say 2000 Hz which is subjectively most important. To prevent the loss of insulation at this frequency the partition should be made less stiff (but without reducing its weight appreciably) so as to move the coincidence frequency up to, say, 4000 Hz which will not usually be subjectively so important.

We have so far been considering only single walls. Double-leaf walls can be used to get more insulation than would be got

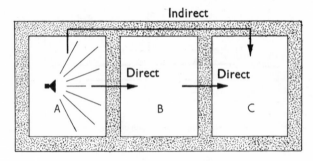

FIG. 63. Insulation between Three Rooms

from their weight alone, but there are limitations at the lower frequencies and it is usually only at the mid and higher frequencies that double-leaf constructions are better than single-leaf (of the same total weight). It depends on the type of noise being insulated against and on the order of insulation whether or not improvement only at mid and higher frequencies is important. It certainly is useful when the average insulation is fairly low, say round about 35 to 40 dB, and when speech is the noise. Thus, partitions between offices might usefully be double. On the other hand, for higher insulations, say 45–50 dB, the higher-frequency noises are not heard through the wall or floor. Thus there would be little point in using double constructions here.

It is sometimes of importance to consider noise transmission from one room to another not adjacent to it. Fig. 63 shows three rooms in a row. A is the source room. If the insulation between A and B and between B and C is low, i.e. so that indirect transmission is negligible, then the total insulation between A and C will be the sum of A to B plus B to C. However, in many cases the indirect transmission will not be negligible and then the insulation A to C will be less than the sum of the two partitions. It is not possible to say in general what the total insulation will be, but as a very rough guide the insulation A to C will be about 10 dB higher than A to B, and so on.

We have so far been considering single or double solid walls (or floors) between rooms. (By solid is meant non-porous and with no holes through it. Walls made of hollow blocks, for example, we would still call solid walls in this context; their insulation would not be any different from solid walls except that the superficial weight would be less.) If a wall has air-paths through it, for example a wall made of clinker concrete blocks, then sound is readily transmitted through these air-paths and the insulation is very much less than would be expected from the mass of the wall. The insulation can be restored by sealing one side—or safer, both sides—of the wall, e.g. by plastering.

A most important aspect of sound insulation—often overlooked—is that the total sound insulation of a composite construction is determined mainly by its weakest link. For example, a 110 mm brick wall between two rooms will have an average insulation of 45 dB. If this wall has built into it a window of a quarter the total area of the wall, the total insulation will be 26 dB (assuming that the window has an average insulation of 20 dB). The method of working out such a combined insulation is given in Chapter 9. A more extreme example of the loss of insulation due to 'weak links' is a 2·5 cm square hole in a 220 mm brick wall, area 62,500 cm² (i.e. about 2·4 × 2·4 m). The 50 dB average insulation of the brick wall will be reduced to 40 dB. This emphasises the great importance of sealing all air-cracks if good sound insulation is to be achieved.

Doors are often a weak link. For one thing, their superficial weight is usually less than that of the wall they are built into; for another thing, the air-cracks round them offer an easy passage for sound unless they are sealed.

A not so obvious example of a weak link is when partitions are carried up only to the underside of a perforated false ceiling. The noise in one room will get through the perforations into the space between the top of the ceiling and the bottom of the structural floor, and will then spread throughout this space. As

there will usually be some sound-absorbent such as glass wool
on the top of the ceiling this space will be rather like an
acoustically-treated duct, and the sound will be attenuated as
it travels by something like 3 dB per metre. But the sound will
only have a metre or so to go to get over the partition and will
get through the ceiling into the next room.

One interesting point about this sort of weak link is that it is
difficult to detect by listening to speech through the partitions.
This is because of the Haas effect (see p. 141), which, stated
briefly, is that if speech is heard from two sources the speech
that arrives first determines the apparent direction. In this case,
a listener on one side of the partition listening to someone speaking on the other side will hear first the sound that comes
through the partition followed a little later by the sound that
has travelled via the ceilings. Even if this second sound is as
much as 10 dB louder than the first sound because the path via
the ceiling has less insulation than the partition, it will appear
to the listener as if all the sound is coming through the partition
and that the partition insulation is 10 dB worse than it really is.
This effect disappears if a steady noise, instead of a voice, is
used.

How much insulation is required depends entirely on the conditions but we should mention here that it is bound up with the
background noise level. For example, between two offices both
used for typing the insulation need not be more than 20 dB and
this will be sufficient to reduce the typing noises from the next
room sufficiently to be negligible. But if one of the rooms is for
typing and the other is a private office then some 40 dB would
be necessary for the typing noise to be made negligible in the
private office.

Another form of background noise may be getting into the
rooms from outside, e.g. traffic noise. Thus traffic noise getting
into a private office might not be sufficient to detract the
occupant but might yet be sufficient to mask typing noises from
an adjacent office which would otherwise be objectionable.
Effective insulation is also affected by the amount of sound
absorption present. If this private office had an absorbent ceiling
then it would reduce the traffic noise about as much as it
reduced the typing noise from the next room; the net gain is
thus zero. But if an absorbent ceiling is installed in the typing
office then less typing noise will be heard in the private office
and an effective gain in insulation is achieved.

We now come to impact sound insulation, and the only

practically important case is impacts on the floor, e.g. footsteps. The loudness of such impact sounds in the receiving room, i.e. the room below the floor, will depend on the general construction of the floor and particularly on the floor surface. By far the best defence is to reduce the amount of impact energy getting into the floor itself, for example by a resilient surface such as carpet or rubber or cork. Such surface layers are not always practicable (and incidentally do not provide any extra air-borne sound insulation), so instead the floating floor construction is often used. This consists of a concrete screed or a wood raft laid on a resilient layer, usually of glass wool or similar materials, on top of the main structural floor (see Fig. 64). This

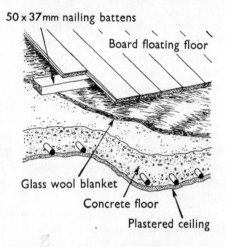

50 x 37mm nailing battens

Board floating floor

Glass wool blanket

Concrete floor

Plastered ceiling

FIG. 64. Concrete Floor with Floating
Wood Raft

construction also improves the air-borne sound insulation. Similar considerations as to the effect of sound absorption in the rooms and indirect transmission apply to impact sound insulation as for air-borne sound insulation.

NOISE TRANSMISSION INTO AND OUT OF BUILDINGS

There may be a direct path for the sound into or out of the building, e.g. an open window, or there may be no direct path when it will be a problem of sound insulation. Consider first noise from outside to inside via an opening. At a position inside the building but very close to the opening the noise level will be almost the same as if there were no building there. But at positions away from the opening the noise level will depend on the area of the opening in relation to the area of absorbent in the room: the smaller the opening or the greater the amount of absorption present the less the noise level. When there is no opening the noise reduction from outside to inside will depend on the composite sound insulation of the room. By composite we mean the sum of all the various paths. For example, traffic noise may enter a room facing a street through the windows, the walls, the roof (if it is also exposed to the noise) and the doors (if any). As for noise from one room to another, the insulation is usually determined by the 'weakest link', which for outside noise is often the windows.

Obviously it is better to have such weak links as windows not facing towards the noise source, if this is possible.

For noise getting from inside a building to the open air the problem is similar. If there is an opening then the noise level immediately outside the opening (if the noise source in the room is not near to the opening) will be the same as the reverberant sound level in the room. At positions outside further away from the opening the noise level will depend on the area of the opening and the distance from it; the smaller the opening or the greater the distance the less the noise level. If there are no openings the sound insulation of the room structure will determine the outside noise level.

Enclosing completely a noise source will nearly always give a very much greater noise reduction in the noise level outside than merely screening the source, provided an adequate amount of sound absorption is introduced into the room thus formed, and provided any doors or windows are adequately dealt with.

VENTILATION PLANT

Ventilation plant may cause noise in rooms due either to the aerodynamic noise made by the fan or to the transmission along the ducts of external noise. Also, ducts common to more than one room may provide paths for noise transmission from one room to another.

Noise will travel long distances along ducts with little reduction in intensity unless special measures are taken. The most common method of reducing the transmission is to line the duct with absorbent. The reduction obtained will depend on the lining material and the size of the duct (see Chapter 9, p. 277): in general the greater the absorption coefficient of the lining and the greater the ratio of perimeter to area, the greater is the attenuation. For large ducts it is necessary to use splitters along the length, i.e. impervious sheets lined on both sides with absorbent. The effect is to increase the ratio of perimeter to area.

Obviously any duct treatment should be placed in the system so as to reduce both the fan noise and the external noise, i.e. it should come after the fan and not before it. It should also be remembered that noise will enter via the outlet duct as readily as via the inlet duct.

When a duct opens into a room the noise level close to the duct exit (i.e. the 'direct sound') is usually much greater (of the order of 20 dB) than the reverberant sound field in the room due to the duct noise. Thus in any room used for speech or music the duct openings should not be close to any of the listeners.

Ducts are often made of thin sheet metal which has little sound insulation. Thus in some circumstances the attenuation along a lined duct may be by-passed by transmission through the sides (see p. 230).

VIBRATION ISOLATION

The isolation of vibration will now be considered. First, we should distinguish what we mean by 'vibration' vis-à-vis 'noise'. By 'vibration' we mean the movement to and fro of a structure, part of a structure or any other solid body caused by some alternating force, such as an out-of-balance rotating piece of machinery. This alternating force may cause the source to vibrate (i.e. the machine to vibrate if it is a machine) or it may cause the structure to which the machine is fixed to vibrate, and in turn it may cause some other part of the same structure to vibrate because of forces transmitted through the structure. From this point of view the ground is considered part of the structure, and thus a vibrating machine in one building may cause vibrations in another building because of transmission through the ground.

Vibration may have four effects. First, it may cause damage; secondly it may be annoying to the occupants; thirdly it may interfere with work, e.g. precision instruments; and lastly it may cause noise.

It can be said at once that damage due to vibration is rare, and long before the vibration becomes damaging it will be intolerable to the inhabitants. For example, an amplitude of vibration (i.e. maximum displacement from the mean position) of a floor of 0·05 mm at a frequency of 10 Hz would be annoying to a person standing on the floor, but it would have to be at least twice that amplitude to cause minor damage. It should also be remembered that much minor building damage, e.g. cracking of plaster ceilings, always occurs in buildings due to the normal movements of the structure resulting from thermal and moisture changes. These cracks may often not be noticed until the building is being looked at carefully to see if there is any vibration damage.

Any source of vibration at an audio frequency will make a noise as well but this is no different from any other noise. Where a difference often does arise is that the vibrations may travel through the structure and make a noise in some other room. There is no way of estimating what the noise might be, but if in a particular case a noise is being made it will be reduced by an amount equal to or greater than any vibration isolation which may be introduced (see p. 284).

The isolation of complete buildings from vibration does not seem to be practicable. However, individual rooms, e.g. broadcasting studios, might be successfully insulated, for example from the vibrations due to underground trains close to the site. But there is little practical experience available.

We will now consider the mechanism of vibration isolation, and this applies equally whether we are considering the isolation of the vibration source, e.g. a machine, from the building it is in, or the isolation of some object, e.g. a delicate instrument, from a vibration existing in the building, due, say, to traffic. We should point out that we will consider only the simplest form of vibration—with one degree of freedom, i.e. vibration in one direction only. In practice, this is adequate for most kinds of machinery vibration (but may not be adequate when dealing with the isolation of rooms). For example, a drop hammer when it strikes will vibrate vertically, and the vertical vibration is the only important one. On the other hand, some form of hammer machine might deliver blows in the horizontal direction and then the horizontal vibration will be important. There are also more complicated machines which produce vibrations in both directions, or, what comes to the same thing, vibrations in a direction somewhere between horizontal and vertical. However, in practice for machines it is usually the vertical vibration that is important, and the more complicated modes of vibration will not be dealt with here.

Fig. 65 represents a source of vibration mounted on some form of resilient mounting (shown diagrammatically as a spring but in practice it could equally well take some other form) which in turn rests on the floor. The source of vibration will usually have a predominant frequency at which it is vibrating (if there is more than one frequency the lowest one should be considered—the higher ones will then also be taken care of) which is called the driving frequency. The resilient mounting with the weight of the machine on it will have its own

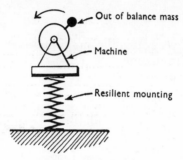

FIG. 65. Illustration of
Vibration Isolation

natural frequency of oscillation, i.e. the frequency at which it will oscillate if given a deflection and then allowed to move on its own. The amount of isolation given by the mounting, i.e. the reduction in transmitted force, which is the same as the reduction in movement of the ground, depends on the ratio of these two frequencies. No isolation will be obtained if the natural frequency of the support is higher than the driving frequency and if they are nearly equal then *more* force will be transmitted, i.e. the resilient mounting will be making things worse. It is only when the natural frequency of the supports is *lower* than the driving frequency that any benefit is obtained. It follows that resilient mountings must be properly chosen; this is discussed on p. 234.

8

Sound Insulation and Noise Control Practice

The first and very often the most important step in designing for sound insulation and noise control is planning. By planning we mean the entire series of preparatory operations from choosing a site, through laying out the buildings, environs and roads, to deciding the position in plan and section for a plumbing duct. It is perhaps unfortunate that in the normal way sound is not represented on a plan. For example, the little rectangle labelled 'diesel generator set' looks so innocuous that there appears no very good reason for not planning a bedroom window a few yards away from it. If, however, the diagrammatic representation of this machine automatically carried with it a great black blot, perhaps some 50 metres or more in diameter, signifying the noise field produced by it, then there would be less chance of this important aspect being overlooked. It is impossible to exaggerate the need to make a close study of all plans, regional, town, site and so on right down to the 1:25 detail of buildings, to ensure that everything possible has been done to *avoid* the demands for highly sound-insulating structures either when the buildings are first erected, or worse still, after they are completed when they have to be installed as 'remedial' measures.

The basic rules are simple; they are:

- (*a*) Separate noise sources from areas requiring quiet by the greatest practical distances.
- (*b*) Plan buildings or rooms not particularly susceptible to noise to act as screens or baffles between noise sources and areas requiring quiet.
- (*c*) Locate the rooms from which noise may originate on any part of the site where there is likely to be noise from other

(exterior) sources. Conversely, locate rooms requiring quiet on a quiet part of the site or side of the building.

(*d*) Locate machinery and any other noise sources which radiate some of their sound by contact with the building low down in the basements if possible. The general structure is likely to be heavier and therefore more sound-insulating at this part of the building and the vibrations can be absorbed into the earth on which the building stands.

(*e*) Remember the loopholes: an open window or a flimsy door in a heavy and otherwise highly sound-insulating wall will bring the overall insulation down to a very low value in spite of the best intentions.

These rules are applicable to all kinds of buildings and developments and will not be restated. We will only state an axiom, which in any case will become self-evident from what follows: the more sound insulation required the more expensive the construction. The rest of this chapter is concerned with the problems associated with different types of buildings and aspects of noise control and will endeavour to give practical answers to these without recourse to more than a minimum of scientific data.

DWELLINGS

In the United Kingdom the Building Regulations 1965 and the Building Standards (Scotland) Regulations 1963 require certain degrees of sound insulation between dwellings. These are broadly based on the grading system for sound insulation recommended by the Building Research Station described in Chapter 10. However it should be noted that constructions giving Grade II insulations do *not* meet the requirements of the regulation nor is the slightly lower standard of party wall insulation called 'Grade I flats' apparently regarded as suitable for flat party walls in England, according to current readings of the Regulations. These (English) Regulations are, in our opinion, not sufficiently explicit in their requirements.

(i) DETACHED HOUSES

The insulation between detached houses will obviously depend on the layout of the adjacent houses and the distance between them. As a rough indication it can be said that for two detached houses separated by a metre or so, with the windows closed and with no windows facing each other directly, the insulation would be of the order of 60 or 70 dB. This is very much higher than the insulation obtainable between non-detached houses.

(ii) PARTY WALLS BETWEEN TERRACED OR SEMI-DETACHED HOUSES AND BETWEEN FLATS

Party wall constructions which meet the recommended standards for sound insulation are given in Table V. Note that Grade I insulation for flats means that noise will be only a minor nuisance, while Grade II means that it will be a major nuisance but still not enough to disturb seriously the majority of the tenants.

TABLE V

PARTY WALLS BETWEEN HOUSES AND FLATS

(a) *Suitable for Houses and Grade I Flats*
    220 mm solid brick, plastered
    280 mm cavity brick, plastered
    175 mm dense concrete, plastered
    300 mm no-fines* concrete, plastered
    Two leaves of 75 mm clinker concrete, plastered, with not less than 75 mm cavity (butterfly wall ties)

(b) *Also suitable for Grade I Flats, but not for Houses*
    220 mm no-fines* concrete, plastered

(c) *Suitable for Grade II Flats, but not for Houses*
    150 mm no-fines* or clinker concrete
    110 mm brick, plastered
    Two leaves of 50 mm clinker concrete, plastered, with not less than 25 mm cavity (butterfly wall ties)

*Assumed density of 1600 $\bar{K}$g/m^3.

The constructions given for houses assume that the remainder of the construction is of a traditional type. With non-traditional construction (such as light-framed construction with claddings and linings) the insulation may be more or it may be less because of differences in the flanking transmission. However, it is unlikely that such differences would be important.

Any of the constructions for flats may be built as panels within the bays of normal heavy steel-framed or reinforced-concrete framed buildings without affecting the insulation grading, provided the connection between the frame member and the wall panel is both airtight and rigid by sealing it with mortar. The insertion of strips of non-rigid material (such as cork or felt) round the edges of the panel to separate it from the frame is not recommended; in certain special constructions such strips can be designed to improve the insulation but in general they are likely to reduce it. In particular, horizontal resilient membranes alone inserted in the walls at floor levels—usually with the object of reducing the flanking transmission up and down the walls— are of no value for the purpose and are best omitted.

A solid party wall made of any other material than those listed in Table V will also give satisfactory insulation for houses and Grade I insulation for flats provided that its superficial weight is at least 440 Kg/m² and provided that there are no air-paths through it. If the material is porous (and this applies of course to no-fines concrete and to clinker concrete) then it is essential that the air-paths through the material should be sealed by plastering and that any flues should be lined or rendered. The lining of such porous walls with wallboard on battens in lieu of plastering is *not* a sufficient seal. Another un-satisfactory construction is dense concrete cast in permanent wood-wool slab shuttering, with the surface of the wood-wool plastered. This has the effect of reducing the insulation—at an important part of the frequency range—to less than it would have been without the wood-wool slabs. A cavity party wall of two leaves of 110 mm brick will be satisfactory provided the cavity is at least 50 mm and only butterfly wall-ties are used. Otherwise, the insulation will be less than that of the 220 mm solid brick wall. Cavity walls of other materials than brick are satisfactory, with the same provisos and so long as the total superficial weight is not less than 440 kg/m². If, however, a light-weight, porous concrete is used in a cavity wall, then the super-ficial weight can be reduced to 244 kg/m² provided the cavity is at least 75 mm.

Grade II insulation for flats will be given by any sealed, solid wall of weight at least 244 Kg/m².

If semi-detached or detached houses are planned so that, downstairs, only kitchen and hallways are adjacent to the party wall, then the sound insulation could be less. However, at least one bedroom upstairs may be adjacent to the party wall and the full insulation would then be necessary.

(iii) PARTY FLOORS BETWEEN FLATS

Common floor constructions which will give Grade I insula-tion overall (i.e. for both air-borne and impact sound insulation) or Grade II insulation overall are listed in Table VI.

TABLE VI 2

PARTY FLOOR CONSTRUCTIONS

(a) *Grade I Overall*

Concrete floor with a floating screed and any soft surface finish.

Concrete floor with a floating wood raft.

Heavy concrete floor (150 to 175 mm solid slab) with a soft floor finish or covering.

Joist floor with floating raft, lath-and-plaster ceiling and 83 Kg/m² pugging and supported on thick walls.

Concrete floor with suspended ceiling and a soft floor finish or covering.*

Concrete floor with 50 mm lightweight screed with a dense topping and a soft floor finish or covering.*

Joist floor with floating raft, lath-and-plaster ceiling and 14·5 Kg/m² pugging and supported on very thick (i.e. more than 220 mm) walls.*

Joist floor with floating raft, lath-and-plaster ceiling and 83 Kg/m² pugging direct on ceiling and supported on thin (i.e. 110 mm) walls.*

* Probable grading.

(b) *Grade II Overall*

Concrete floor with floor finish of wood boards or 6 mm thick linoleum or cork tiles.

Concrete floor with soft floor finish.

Concrete floor with suspended ceiling and wood board floor finish.

Joist floor with plaster-board and single-coat plaster ceiling and 14·5 Kg/m² pugging on ceiling—THICK WALLS.

Joist floor with heavy lath-and-plaster ceiling (no pugging)—THICK WALLS.

Joist floor with lath-and-plaster ceiling and 83 Kg/m² pugging on ceiling.

Joist floor with floating floor and plaster-board and single-coat plaster ceiling (no pugging)—THICK WALLS.

Joist floor with floating floor and lath-and-plaster ceiling and 14·5 Kg/m² pugging on ceiling.

In the above the following definitions are used:

A concrete floor is a reinforced-concrete slab, concrete and hollow pot slab, or hollow concrete beam slab weighing not less than 220 Kg/m² and not less than 100 mm thick.

A soft floor finish is thick cork tile, or rubber on sponge-rubber underlay, or thick carpet.

A plain joist floor is one employing timber or light metal joists with nailed tongued and grooved board finish unless otherwise stated.

As has been mentioned previously, the sound insulation of joist floors is affected more by indirect transmission than is the sound insulation of concrete floors. Thus the total insulation will depend to a greater extent on the supporting wall construction. In Table VI it is indicated that some of the joist floors must be supported by thick walls if they are to meet the grade, a thick wall system being defined as one in which at least two of the supporting walls are at least 220 mm thick.

The required specifications of floating screeds, wood rafts, light-weight screeds, pugging and suspended ceilings are given later.

These Building Research Station recommendations pose a number of questions which architects must either themselves answer, or be prepared to advise those who will. Decisions will have to be made as to whether any special precautions are to be taken to ensure good sound insulation, and if so, up to what standard. It would seem extremely unwise to adopt methods of construction which are known to produce public protests from the occupants (the so-called 'deputation level' of insulation). This rules out the use of all floors having an insulation much worse than Grade II, viz. all joist floors supported on thin walls and without pugging; also all joist floors supported on thick walls and without pugging unless they have a heavy lath-and-plaster ceiling or a floating floor.

If dwellings are to be erected in very quiet rural or suburban districts it is more important to ensure high values of insulation because noise from the neighbours will be noticed more than in noisier surroundings. This fact may, in part, account for the regional differences in opinion of noise nuisance reported in Social Surveys where the occupants of flats who had probably been accustomed to living in noisy and overcrowded conditions for most of their lives were less concerned about the insulation than those who were accustomed to better standards. This is not to advocate low insulation for town or city dwellings but merely to imply that while a high background noise in towns remains unavoidable, the relative importance of well-insulated dwellings is less.

It may be practical to vary the nature of the party walls in a block of flats depending on the rooms which adjoin them. Such variations in floors would probably not be practical in normal flat planning where all types of rooms occur on one floor level. However, in certain of the more unusual maisonette blocks it may well be possible to reduce the number of high-insulation floors depending on the room dispositions. It is pointed out that if bedrooms occur under living-rooms in a maisonette plan, the conditions are most stringent and Grade I floors should be used in all noise climates.

INTERNAL SOUND INSULATION IN DWELLINGS

We have so far been considering only the insulation between adjacent houses and flats. Very much lower insulation is tolerated between rooms in a house or flat occupied by one family. Nevertheless, it appears that the internal insulation should not be less than about 30 dB. Any floor construction will be adequate, but partitions between rooms will be lower than this unless their superficial weight is 24·5 Kg/m² or more. Thus, for example, a partition between bedrooms consisting only of cupboards is unsuitable. Examples of suitable constructions can be obtained from Table VII, p. 216.

SCHOOLS

For general planning recommendations see the basic rules given at the beginning of this chapter and the section on Noise Reduction by Site Planning, p. 223.

For the particular problem of protecting schools from high outside noise levels use may be made of an acoustically controlled automatic window closing system developed by the Building Research Station. This system was originally applied to a school in the vicinity of London Airport where the extremely high, but not continuous noise from aircraft movements are satisfactorily abated. There is no reason, however, why the scheme should not be used where trouble from loud noise of other sources (for example trains) provided that the total duration of the very loud noise during the working day is not so long that ventilation of the classrooms suffers.

In the lack of any extensive study of the needs of sound insulation in schools it is impossible to make definitive recommendations as has been done for dwellings. The insulation required between the various rooms of schools and other educational buildings will depend on a number of factors. Broadly, rooms may be rated on a system describing their propensities as noise sources and their tolerance of incoming noise. The nomogram Fig. 66 may then be used to obtain some indication of the desirable standards of average sound insulation.

Each room must, of course, be considered both as a noise source and as a noise tolerator. For example, an ordinary classroom might be regarded as an average noise source and an average tolerator; a music classroom as a high-level source and a low tolerator; whereas a carpentry classroom would be a high-level source and have high tolerance of incoming noise.

The stopping off of the insulation scale at the top signifies that it is considered impractical to use a wall of high enough insulation between a room of high noise level and one of low noise

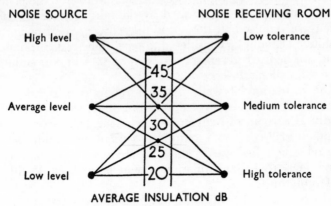

FIG. 66. Suggested Sound Insulation for Schools

tolerance. This problem should be solved by planning the rooms so that they do not adjoin.

These recommendations agree broadly with those given in British Standard Code of Practice CP 3, Chapter III (1960) which was published subsequently to the original edition of this book.

For constructions to give these average values see Table VII, p. 216.

This nomogram should apply both to walls and floors as far as air-borne sound insulation is concerned. However, floors also have to provide impact insulation and it is considered that all floors should have at least Grade II insulation as specified for dwellings above.

To avoid the spread of noise through the building all corridors and circulation spaces should have highly sound-absorbing ceilings, as should dining-rooms and rooms rated as high-level noise sources.

HOSPITALS

A few general recommendations can be given in addition to the basic planning rules at the beginning of this chapter. Large multi-storey hospitals present a more difficult problem than small single-storey ones. The main kitchen should if possible be put in a separate building connected to the wards only by a service lift. If this is impossible it should be planned beneath the wards rather than above them. The importance of reduction of noise at source cannot be overstressed. Many of these noises result from and are emphasised by the hard, hygienic surfaces which are required in these buildings. Every item of hospital equipment should be considered to see if some resilient material could not be used to replace the hard ones. All doors should be fitted with silent closers.

The insulation of walls between rooms in general, except where neither require protection from noise, should be at least 40 dB. If a room of high noise level, such as a ward kitchen, occurs next to a room requiring very quiet conditions, for example single-bed wards which are frequently used for acutely ill patients, then the insulation should be at least 45 dB, and preferably more.

Floors between wards should preferably have an overall insulation of 50 dB or at least 45 dB.

Most of the common sound-absorbing surface treatments are not popular with the medical profession because of fears about their germ-harbouring possibilities. Special treatments such as thin non-porous films over some soft absorbent material—which are now commercially available—might overcome these objections, or research may prove that the normal materials are innocuous. They are extensively used in the United States of America. The use of absorbent surfaces can be most helpful in reducing noise and they should certainly be used in corridors and circulation space. They should not be used too liberally in wards because they can increase the intelligibility to an embarrassing degree.

CONCERT-HALLS, THEATRES, CHURCHES AND OTHER AUDITORIA

In these buildings the main requirements are to reduce all noise from outside and any source inside the building to a very low value in the auditorium. The amount of insulation to be provided obviously depends on the intensities of the noise sources, and again the basic planning rules at the beginning of this chapter should be observed, in conjunction with the section on planning distances on p. 225. It is most desirable for any important building that a noise survey of the site be made so that the insulation required can be accurately determined.

Any important auditorium on a town or city site should have a protective zone of rooms between the outside and the auditorium proper so that there are in effect two walls reducing the noise. This system may also have to be used on the roof if the noise climate is very high. The increase in air traffic makes it desirable that the roof is made as heavy as possible. Although aircraft noise may not last very long it may occur at a quiet passage in the performance and thereby spoil the whole presentation.

Generally speaking windows should be avoided as it is very difficult to provide practical and economic construction which will give the necessary insulation. When windows are essential (and they can safely be used only on fairly quiet sites) they should be as small as possible and should be in fixed sealed frames, single or double glass, depending on the noise climate.

Doors should be heavy, and when they divide the auditorium from a noisy area they must have adequate means of sealing, or better, the entrances be provided with sound locks (see p. 221).

It is not practical to isolate large auditoria from the vibrations and consequent low-frequency noise produced by trains, either on the surface or underground. The best that can be done to reduce the danger is to put the auditorium as high above the ground as possible on columns, and to construct a number of solid floors carried on the same columns between the auditorium floor and the ground, as was done in the Royal Festival Hall (see Fig. 67).

Some practical information on the siting and general arrangement of ventilation systems is given on p. 226.

All circulation spaces, foyers, bars, restaurants and any other rooms in the protective zone must have sound-absorbent ceilings at least, and all floors surrounding the auditorium should have soft finishes.

Music rehearsal rooms should be situated as far as possible from the auditorium. If they were to adjoin, an insulation of at least 70 dB would be necessary and this could only be achieved by very special construction.

STUDIOS:
BROADCASTING, TELEVISION AND RECORDING

The requirements are to reduce all noise from outside and any source inside the building to very low values. These values

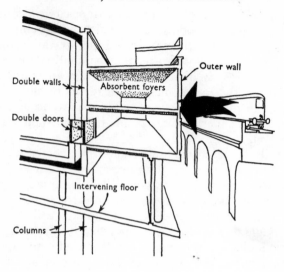

FIG. 67. Section of the Royal Festival Hall showing Sound Insulation
Measures

vary with the room use. The list below indicates the tolerance
to noise and interference from other non-noise sources, i.e.
sounds having an intelligible content.

Room	Rating as Noise Source	Tolerance of Incoming:	
		Noise	Interference
Music studio, radio or recording	High	Very low	Very low
Talks and drama studios, radio or recording	Medium	Very low	Very low
Control and listening rooms, radio, recording or television	High	Low	Very low
Television studios, including dubbing suites	High	Low	Very low
Recording rooms	Medium	Low	Low

The amount of protection required from external noise
depends on the intensities of the noise sources. For busy city or
town sites it is probably essential that rooms with very low
tolerance of noise be provided with a protective zone between

the outside of the building and the rooms, as exemplified in the design of Broadcasting House, London (see Fig. 68). In quieter surroundings these rooms may be protected by a single wall having a weight of not less than 490 Kg/m², and which should be without windows unless these consist of at least three sealed layers of glazing.

Rooms with low tolerance of noise can have windows to the outside. These should be double fixed glazing in town sites or thick single glass fixed windows in quieter sites.

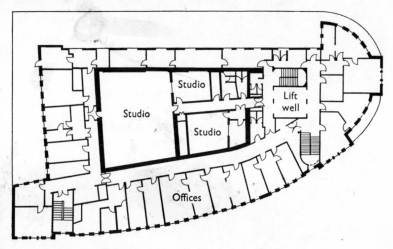

FIG. 68. Plan of Broadcasting House, London

An idea of the necessary insulation between the rooms can also be gained from the above table. For example, a music studio adjoining a talks studio would require a maximum degree of insulation. An overall value of about 75 dB is necessary. This could only be achieved by walls having a total weight of at least 980 Kg/m² and of cavity construction with careful attention to the reduction of indirect transmission. The best solution is to plan such rooms apart, and they should certainly never be planned one above the other with only a single-floor separation. On the other hand two talks studios could adjoin if they were separated by a wall having a weight of not less than 490 Kg/m² and careful attention is given to reduction of indirect transmission.

The overall insulation between a studio and its control room should be not less than 45 dB average. The observation window which is normally provided may have an insulation slightly less than this and therefore a 50-dB wall should be used. A detail of the recommended construction of an observation window is given in Fig. 69.

Walls between studios or control rooms and circulation space (which should be of limited use for access to the technical areas only) should be at least 50 dB for studios and 45 dB for control rooms. Such circulation areas *must* have sound-absorbent ceilings at least (upper part of walls also is preferable) and soft floor coverings.

All doors should be provided with sound locks, or lobbies, which should be acoustically treated as recommended for

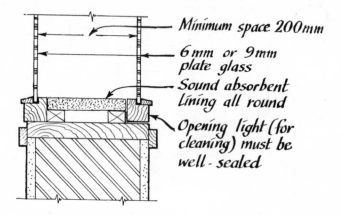

Minimum space 200mm

6mm or 9mm plate glass

Sound absorbent lining all round

Opening light (for cleaning) must be well-sealed

FIG. 69. Detail of Observation Window for Studios

general circulation areas above. By using the principle of two doors and a sound lock between all rooms needing high insulation the need for very elaborate and unwieldy sound-insulating doors is avoided. However, the doors should not be less heavy than those shown detailed in Fig. 70, and must have proper sealing arrangements round their edges.

Where rooms for general occupation occur over studios these should have 50 dB insulation and soft floor coverings, and the studios should have heavy suspended ceilings. A section of a typical studio suite in a normal steel-framed office-type building is given in Fig. 71.

OFFICE AND INDUSTRIAL BUILDINGS

In multi-storey buildings primarily for office use the main problems are protection from external street noise, and adequate

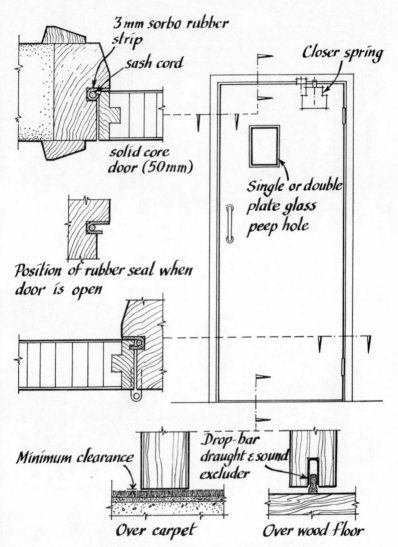

3 mm sorbo rubber
strip

sash cord

Closer spring

solid core
door (50mm)

Single or double
plate glass
peep hole

Position of rubber seal when
door is open

Minimum clearance

Drop-bar
draught & sound
excluder

Over carpet

Over wood floor

FIG. 70. Detail of Studio Door

insulation between the individual rooms. The basic rules of planning should be followed, but where rooms requiring especially low noise levels, such as board rooms and conference rooms, must face noisy streets double glazing or single sealed plate-glass windows to provide a minimum of 30 dB insulation are essential. The mainly glass curtain wall may be satisfactory

203

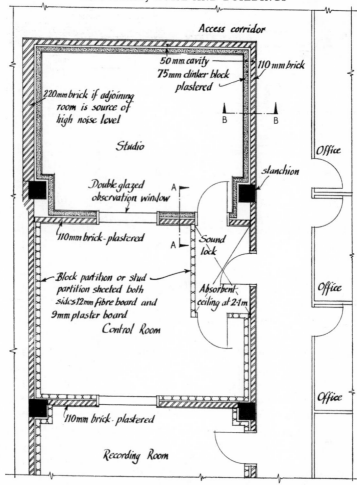

PLAN
FIG. 71. Construction of Studios in a Steel-framed Building
(Sections on facing page)

for some rooms but it is likely to cause complaints on very noisy
sites, particularly on the lower floors of the building, if quiet
office conditions are sought.

The division of lettable office space by demountable dry-
construction office partitioning is becoming increasingly common. Most of these systems provide a measured overall sound
insulation between offices of between 20 and 25 dB. This value
of insulation will often be sufficient in noisy locations where the
background of traffic noise masks the sounds coming through
the partitions. It is stressed that the presence of background

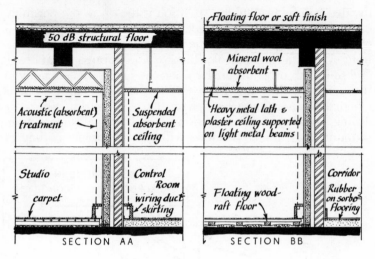

SECTION AA

SECTION BB

noise from traffic makes a great difference to the practical effectiveness of partitions as sound insulators. The nomogram of Fig. 72 is suggested as a guide to design aims. In using this, the following definitions should be observed. Under 'background noise condition' use 'quiet' column for rural, suburban and city sites well away from traffic or where double windows are installed to reduce traffic noise. Use 'noisy' column for sites where offices face on to busy streets. A quiet office is defined as one for executive or administrative staff occupation where a high degree of privacy is required. An average office is one for general clerical staff, equipped with typewriters or a few of the quieter types of accounting machines. A noisy office would be a typists' pool, or room for addressograph or other noisy accounting machines. A sound-absorbent ceiling should always be used in these rooms.

To obtain insulation of 30 dB or more partitions built in normal building techniques may be necessary (see Table VII, p. 216), or it may be possible to obtain at least 30 dB from a well-designed demountable system. Most of the systems tend to show an overall insulation of about 5 dB less than the mass value, probably due to inattention to proper sealing of cracks in the construction. Exaggerated claims of insulation by some of the manufacturers of partition systems have been frequently seen.

Where partitions are erected leaving a gap between the top and the ceiling these should be regarded as having a space-defining function only. The sound insulation between rooms so divided will be negligible. Attention is drawn to the possibility of loss of insulation where by-pass routes exist round the ends or over the top of partitions. One example is where partitions are built up to light suspended ceilings (e.g. perforated metal heated and sound-absorbent ceilings) when the sound passes readily up through the ceiling and down on the other side. This loss can be avoided by inserting a barrier on the partition line above the ceiling. Another is where partitions are built against continuous radiator ducts under windows without any proper sealing off

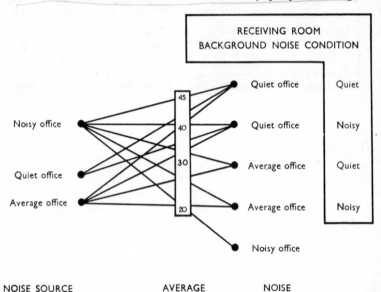

Fig. 72. Suggested Sound Insulation for Offices

of the space at the line of the partition. Again, a barrier must be inserted across the duct made good against all pipes. Overall insulation can be reduced to as low as 10 to 15 dB by these defects unless they are properly dealt with.

All rooms which constitute a noise source such as accounting-machine rooms, typists' pools, etc., should be equipped with sound-absorbent ceilings. It is also very desirable to similarly treat corridors and offices generally.

When offices are incorporated in factory or other industrial buildings there may be a need to protect them from the factory noise. A method of assessing the required degree of insulation is described on p. 290.

Further information on the general problems of noise in industry is given on p. 235.

DOCTORS' CONSULTING-ROOMS

The problem of overhearing between a doctor's consulting-room and an adjoining waiting-room is one which is often met. To ensure absolute privacy, bearing in mind the conditions of a low noise climate and keen interest on the part of some of the audience in the waiting-room, an overall insulation of at least 45 dB is essential. The dividing wall must therefore be of 110 mm brickwork or some other construction giving the same insulation, but it will be seen from Table VII that it will be inadmissible to have any kind of single door in the wall. Two 50 mm solid wood doors in conjunction with a sound lock are necessary if they are directly in the wall dividing the rooms. Two lighter flush panel doors would suffice if they were planned in conjunction with a connecting corridor which would form a large sound lock.

AUDIOMETRIC ROOMS

Rooms for the most exact forms of audiometry and for some other scientific work demand the very highest degree of sound insulation. If the room can be in a separate building located in a low noise climate, then a solid and heavy construction will usually suffice. However, when such rooms are contained in buildings of general occupation, the completely isolated box type of structure is the only one which will be likely to provide the necessary defence against external noise.

The illustration Fig. 73 shows the general nature of the constructional measures necessary. Rooms with this type of construction could safely be located next to any rooms except ones containing exceptionally high noise levels.

CONSTRUCTIONAL DETAILS OF WALLS AND FLOORS

Constructional details of some of the basic sound-insulation measures referred to above and in the Tables V and VI are given below.

CONSTRUCTION OF CONCRETE FLOORS

Wood-raft Floating Floors (on Concrete)

The wood-raft floating floor consists simply of floor-boarding nailed to battens to form a raft which rests on a resilient quilt laid over the structural floor slab. The battens must not be fixed in any way to the slab. For structural reasons tongued and

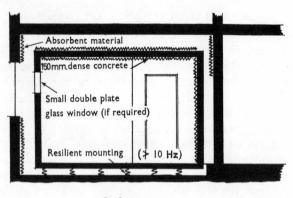

Side view

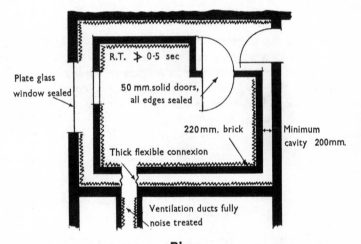

Plan

Fig. 73. Detail of an Audiometric or other Highly Insulated Room

grooved floor boards are preferred not less than 22 mm thick. A section of the construction is shown in Fig. 74.

The resilient layer is a very important feature of the floating floor. Glass wool and mineral wool are best for resilient layers, quilts of the long-fibre type being the most satisfactory. Other materials are sometimes suggested and are occasionally used, but at present no alternative is known of comparable cost and general suitability. A nominal quilt thickness of 25 mm at a density of 48–96 kg/m³ is recommended; it compresses to about 9 mm under the battens of a wood-raft floor. Thicker quilts give better insulation but allow too much movement of the floating floor. Bitumen-bonded mats are the cheapest form of quilt and they are sometimes used on this account, but paper-covered plain wool quilts are preferred because of their better performance for impact sound insulation and because they are less

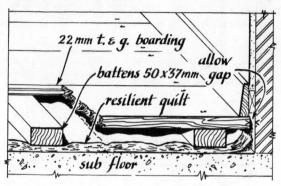

FIG. 74. Section of a Wood-raft Floating Floor

susceptible to damage from the slight rocking movement of the raft battens when the floor is walked on. Most resilient quilts are obtained in rolls 1 m wide. They should be laid with edges closely butted but not overlapping.

Softwood is generally used for floating floors in dwellings; if hardwood is used in other buildings special precautions are necessary because hardwood is usually supplied kiln-dried to a low moisture content. If this moisture content is lower than it will subsequently be when the building is completed and in use, the wood will swell as it takes up moisture and the floating raft, being unrestrained, will buckle. Therefore when using hardwood for floating rafts it is desirable to construct the raft with the hardwood at a higher moisture content than that at which it is usually supplied, and, if possible, to use types of hardwood with a comparatively low moisture movement.

Concrete-screed Floating Floors

This type of floating floor consists of a layer of concrete not less than 38 mm thick (50 mm thick or more is better) resting on a resilient quilt laid over the structural floor slab and turned up against the surrounding walls at all edges, as shown in Fig. 75. As with wood-raft floating floors the best resilient quilt is glass wool or mineral wool, of 25 mm nominal thickness at a density of 48–96 kg/m³. The quilt compresses to 10–12 mm under the load of a floating screed. Paper-covered quilts of plain wool give better insulation than bitumen-bonded mats; nevertheless, the latter have been extensively used under floating concrete screeds because they are cheaper, and in many instances have proved satisfactory. It is necessary to provide a layer of waterproof building paper over the quilt to prevent wet concrete running through it. If the quilt is supplied with its own covering

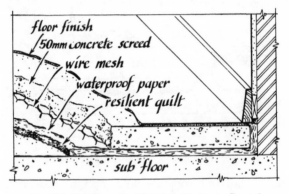

FIG. 75. Section of a Floating Concrete Screed

of waterproof paper, it may be sufficient to provide extra waterproof paper in narrow strips (say 150 mm wide) covering the joints in the quilting only. It is recommended to provide wire-mesh reinforcement (e.g. 20–50 mm mesh chicken wire) for the floating screed, laid directly on to the waterproof paper before placing the screed. Apart from reducing the risk of cracking, the wire netting protects the building paper and the resilient quilt from mechanical damage during the operation of placing the concrete; any such damage might well result in the concrete making direct contact with the structural slab, which would spoil the insulation of the floating floor. A suitable mix for the concrete screed is 1:2:4, cement:sand:gravel aggregate, with an aggregate size of not more than 10 mm.

Concrete-screed floating floors may not be suitable for rooms larger than about 15 m² or 20 m² at most, or with a greater dimension than 5–7 m. This is because a slight 'dishing' or curling up at the corners invariably occurs with floating floors, owing to quicker drying out at the top surface than at the bottom and to the lack of bond with the sub-floor. This curling is not noticeable in small rooms, but in rooms of larger dimensions than those given the risk of cracking from drying shrinkage makes it necessary to sub-divide the floating screed into bays, and the curling of these separate bays can lead to difficulty with many types of floor finish. The foregoing remarks apply to screeds using Portland cement. By employing one of the new cementitious materials such as synthetic anhydrite it may be possible to lay much larger areas.

Floor Finishes and Coverings

In present-day buildings hard floor finishes, such as thermoplastic tiles, pitchmastic and magnesium oxychloride, are commonly used. Finishes of this type make virtually no contribution to the sound insulation of a floor. Ordinarily, no floor finish adds anything to the air-borne sound insulation, but the impact sound insulation can be considerably improved by means of a finish that is sufficiently soft or resilient. Thus it is possible to raise the impact insulation of all floors simply by adding a finish or covering that is soft enough. Thick fitted carpet or underfelt, for instance, will nearly always give Grade I impact insulation for flats. Other floor finishes giving improved impact insulation when applied to plain concrete floor include rubber flooring on a sponge-rubber underlay or thick cork tiles. A table of comparative impact insulation of different floors and finishes is given on p. 219.

Partitions on Floating Floors

Partitions (apart from such short lengths as the sides of cupboards) should not be built on top of floating floors, which should be self-contained within each room. Building the partitions on floating floors may overload the resilient quilt and reduce its insulating properties; moreover the movement of the floating floor may cause the partitions to crack.

Pipes and Conduits in Floating Floors

It is often necessary for services such as electric conduits, gas and water-pipes, etc., to traverse a concrete floor. Whenever possible these pipes should be accommodated within the thickness of the floor slab or the levelling screed, but sometimes they have to be laid on top of the slab and contained within the depth of a floating floor. This is more likely to occur with a floating concrete screed. A floating wood raft requires a very level surface as a base and therefore, as a rule, calls for a levelling screed—in which pipes can be embedded. However, pipes or conduits need not cause trouble with a floating screed provided: (*a*) they do not extend more than about 25 mm above the base; (*b*) they are properly fixed so as not to move whilst the floating floor is being laid; and (*c*) they are haunched up with mortar on each side to give continuous support to the resilient quilt. When two pipes cross, one of them should be sunk into the base slab; the pipes can of course be readily accommodated parallel to and between the raft battens, but in the other direction the battens will have to be notched over them.

LIGHTWEIGHT CONCRETE SCREEDS

A lightweight concrete screed sandwiched between the floor and the structural floor slab (without a resilient quilt) often gives 50 dB *air-borne* sound insulation, but the impact insulation is no better than that of a concrete floor with the same floor finish. A soft floor finish such as one of those already described is therefore necessary in addition to a lightweight concrete screed if impact insulation is important. The method is then comparable for sound insulation to the use of a heavy (150 mm to 175 mm) concrete floor but gives a saving in weight. The essential requirements would appear to be as follows:

(*a*) The density of the lightweight screed should be not more than 1120 Kg/m^3.

(*b*) The thickness of the lightweight screed should be not less than 50 mm.

(*c*) An impervious or airtight layer should be provided above the lightweight screed.

The provision of the airtight sealing layer is important. A dense concrete topping is usually required above a lightweight screed to ensure a satisfactory base for the floor finish, and this topping (which should be additional to the 50 mm minimum thickness of the lightweight screed itself) will provide the necessary seal for the screed. Wood boards nailed direct to the lightweight screed are not a satisfactory solution, presumably because they are not sufficiently airtight, even when tongued and grooved.

Suspended Ceilings

Suspended ceilings are chiefly of benefit against air-borne sound and are comparable with lightweight screeds in that they can be used to raise the air-borne sound insulation of a normal concrete floor up to about 50 dB. A soft floor finish would be necessary to give an improvement in impact insulation. Not all suspended ceilings give a satisfactory improvement of sound insulation. The requirements for a successful system of construction are not all known precisely, but the following features would appear to be of importance:

(a) The ceiling membrane should be moderately heavy, say not less than 24 Kg/m^2.

(b) The membrane should not be too rigid.

(c) The ceiling should be essentially airtight so as to eliminate direct sound penetration via air-paths, such as would occur with open-textured materials or with unsealed joints.

(d) The points of suspension from the floor structure should be as few and as flexible as possible.

However, as with wood-wool used as permanent shuttering it is possible for a suspended ceiling to reduce the insulation to below what it would have been without it. This defect appears to occur when the ceiling is rigid, weighs less than 24 Kg/m^2 and when it is attached at close intervals to the structural floor, e.g. plasterboard nailed to battens fixed firmly to the underside of the floor.

In spite of their usefulness for sound absorption, ceilings of soft insulating fibreboard alone are not suitable for sound insulation because of their light weight and porous nature. Plastering on expanded metal or on ceiling boards, provided the total weight is not less than the 24 Kg/m^2 mentioned above, and provided the whole membrane is supported by light metal hangers, is usually satisfactory. The air space above the ceiling may range in depth from 25 mm to 300 mm or more—the deeper the better.

Construction of Joist Floors with Floating Finish

The floating floor consists of floor-boarding nailed to battens to form a raft, which rests on a resilient quilt draped over the joists. The raft must not be nailed down to the joists at any point and it must be isolated from the surrounding walls, either by turning up the resilient quilt at the edges (which is the better practice) or by leaving a gap round the edges to be covered by the skirting. The floor-boards should be not less than 22 mm thick, preferably tongued and grooved. The battens should be 50 mm wide and at least 37 mm deep, preferably 50 mm deep. They should be parallel with the joists because battens that cross the joists provide too small a bearing area and overload the resilient quilt; the whole area of the top edges of the joists should share the loads transmitted from the floating raft—see detail Fig. 76.

There are two common methods of constructing the raft: one

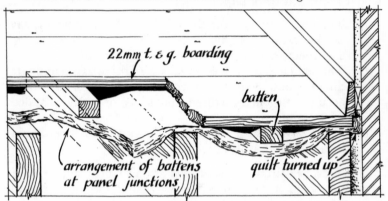

Fig. 76. Section of a Floating Floor on Joists

is to place the battens on the quilt along the top of each joist and to nail down the boards to the battens in the normal way. The other method is to prefabricate the raft in separate panels as long as the room and 0·6 to 1·0 m wide, with the battens across the panels positioned so that they will lie between the joists when the panels are placed into position. The battens should be slightly staggered and project a few inches beyond the sides of the panels in order that the panels can be screwed together to form a complete raft.

Because the flooring is not nailed down particular care must be taken to level up the joists that are to carry a floating floor. In particular, end joists must not be lower than the others as this produces a tendency for the floating raft to tip and for the furniture next to the walls to rock.

RESILIENT LAYERS

The requirements regarding resilient quilts for floating wood-joist floors are no different from those already described for concrete floors. It is good practice to turn up the quilt all round the edges against the walls, though this is not essential if the floating floor does not touch the walls. The gap is normally covered by a skirting fixed to the wall only.

The lath-and-plaster ceiling referred to in Table VI should be of expanded-metal lath and three-coat plaster.

The pugging used in any joist floor should be placed directly on the ceiling and not on separate pugging boards. The 83 kg/m² pugging referred to in Table VI would normally be 50 mm of sand, but any other loose granular material of at least the same weight will do. The material must not be omitted from the narrow spaces between the end joists and the walls. It should be as dry as possible when placed in the floor, and sand containing deliquescent salts must of course be avoided.

The 14·5 kg/m² pugging referred to in Table VI should normally be a 75 mm thickness of high-density slag-wool (190–230 kg/m³), although other materials can be used provided they are of loose wool or granular type and have the requisite weight. Lighter-weight materials such as glass-wool or exfoliated vermiculite are not suitable.

Partitions on Floating Floors of Joist Construction

Partitions should be supported either on the partitions below or on the floor joists, the floating floor being constructed as a separate independent raft within the confines of each room.

Suspended Ceilings and Joist Floors

An independent joist-supported ceiling is not normally recommended as a means of improving the sound insulation of a wood-joist floor, because it is not effective for the purpose. When used in addition to a floating floor and pugging a suspended ceiling produces little further improvement. Used alone, say for improving the sound insulation of an existing wood-joist floor, a suspended ceiling to be of much value needs to be so heavy that it might not be practicable to construct it. It is true that the airborne insulation at *high* frequencies could be improved by a comparatively light suspended ceiling, but wood-joist floors are deficient in sound insulation mainly at the lower frequencies and more weight is required to remedy this deficiency.

Suspended ceilings are not of much benefit for impact insulation. If a floating floor is being built it is usually a simple matter to pug the ceiling, and nothing worth while is then gained by adding a suspended ceiling. If, however, an existing floor cannot be disturbed and carpeting or some other soft floor covering is being added to give improved impact insulation, then the addition of a heavy suspended ceiling also may give a moderate improvement of the air-borne sound insulation, but such a floor is unlikely to approach 45 dB air-borne sound insulation unless the supporting walls are thick.

TABLE VII

TABLE OF AVERAGE AIR-BORNE SOUND INSULATION

This table indicates the air-borne sound insulation, averaged over the frequency range 100 to 3150 Hz, of a number of common types of wall and floor constructions, windows and doors. It is assumed that there are no holes or cracks in the constructions, except those specifically mentioned.

As has been pointed out in Chapter 7 these single-figure values must be taken only as a guide because insulation effectiveness depends on how the insulation varies with frequency and because differences in building construction affect the values actually obtained.

It must also be remembered that the insulation achieved in practice depends not only on the insulation of the particular dividing element but also on its area in relation to the sound absorption in the rooms, and on indirect transmission. No specific allowance can be made for indirect transmission, except to say that for elements having an insulation of 40 dB or below it will have little effect. In the following tables the figures above 40 dB allow for the amount of indirect transmission likely to be present when the structures are used in a more or less traditional manner.

As to the effects of area and absorption, the values given have been chosen to represent as nearly as possible the achieved insulation between two normally furnished rooms of average proportions when the whole area of the wall or floor is of the specified construction.

The above considerations have led to the adoption of 5-dB steps in the presentation of values. To go to finer divisions than this might give a spurious idea of the accuracy to be expected. Estimates for other solid, sealed wall or floor constructions which do not appear in this table can be made from the 'mass law' curve, Fig. 61. A table of the weights of various common building materials is given in Appendix B.

WALLS

About 55 dB
1.   450 mm solid brick or stone.

About 50 dB
2.   220 mm solid brick or stone, plastered.
3.   175 mm dense concrete.
4.   300 mm no-fines concrete, plastered.
5.   Two leaves 75 mm clinker concrete, plastered, with not less than 75 mm cavity—'butterfly' or no wall ties.

WALLS

About 45 dB

6.   110 mm solid brick, plastered.
7.   100 mm dense concrete.
8.   150 mm no-fines concrete, plastered.
9.   Two leaves 50 mm clinker concrete, plastered, with not less than 25 mm cavity, 'butterfly' wall ties.

About 40 dB

10.   75 mm clinker concrete, plastered both sides.
11.   50 mm dense concrete.

About 35 dB

12.   50 mm clinker concrete, plastered both sides.
13.   63 mm hollow clay block, plastered both sides.
14.   Lath-and-plaster (3 coats) both sides of 100 mm studs.

About 30 dB

15.   6 mm plywood or hardboard both sides 63 mm timber studs, with 50 mm glass-wool in cavities.
16.   10 mm plaster board (skim coated) both sides 100 mm studs.

About 25 dB

17.   10 mm plasterboard (skim coated) on one side of timber frame.

About 20 dB

18.   12 mm fibreboard on one or both sides of timber frame.

WINDOWS AND DOORS

About 45 dB

19.   Two 50 mm solid wood doors with all cracks adequately sealed in conjunction with a sound lock.

About 40 dB

20.   Double windows of 3 or 4 mm glass, spacing 200 mm, tightly sealed with absorbent in reveals.
(Better insulation at low frequencies if 6 mm plate glass is used.)

About 35 dB

21.   Double windows of 3 or 4 mm glass, spacing 100 mm, tightly sealed with absorbent in reveals.
(Better insulation at low frequencies if 6 mm plate glass is used.)
22.   Double windows of 3 or 4 mm glass in wood or metal frames spaced 200 mm apart, with opening lights closed but not sealed; absorbent in reveals.
23.   Two flush panel doors (hollow core with 3 mm hardboard or plywood both sides), with cracks adequately sealed and in conjunction with a sound lock.

About 30 dB

24.   Single 6 mm plate-glass window, all edges sealed.
25.   50 mm solid wood door with all edges adequately sealed.

About 25 dB

26.   Single 3 or 4 mm glass window, all edges sealed.
27.   50 mm solid door with normal cracks round edges.

WINDOWS AND DOORS

About 20 dB

28.	Single 3 or 4 mm glass window, wood or metal frames, with opening lights closed in normal manner; not sealed.
29.	Flush panel door (hollow core with 3 mm hardboard both sides), with cracks adequately sealed.

About 15 dB

30.	Flush panel door (hollow core with 3 mm hardboard both sides), with normal cracks round edges.

Notes:

(a) By 'doors with cracks adequately sealed' is meant one of the treatments described on p. 221.

(b) By 'doors with a sound lock' is meant the construction described on p. 221.

(c) The effect on overall insulation of the insertion into a wall of a door or window of lower insulation can be determined from the graph Fig. 93.

Assuming that the wall has an insulation at least 15 dB greater than the door or window, a rough guide to the overall insulation is as follows:

10 dB more than the door or window if its area is not more than one-tenth of the total area.

5 dB more than the door or window if its area is not more than one-third of the total area.

(d) These estimates and those given by the graph, Fig. 93, refer to insulation of walls between rooms. For insulation between a room and a noise in the open air the overall insulation result should be *reduced* by 5 dB.

FLOORS

About 50 dB

31.	175 mm solid dense concrete slab plastered on soffit and any floor finish.
32.	Concrete floor plastered on soffit, with floating wood raft or floating screed floor finish.
33.	Concrete floor with heavy freely suspended ceiling and any floor finish.
34.	Concrete floor plastered on soffit, with 50 mm lightweight screed with dense topping.
35.	Joist floor with floating raft finish, lath-and-plaster ceiling and 50 mm sand pugging directly on ceiling, supported on thick walls.

About 45 dB

36.	Concrete floor plastered on soffit and any floor finish.
37.	Joist floor with t. and g. boards, lath-and-plaster ceiling and 50 mm sand pugging directly on ceiling.
38.	Joist floor with floating raft finish, lath-and-plaster ceiling and 75 mm rock-wool (or similar) pugging directly on ceiling, supported on thick walls.

About 40 dB

39.	Joist floor with t. and g. boards, 10 mm plasterboard and skim-coat ceiling and 75 mm rock-wool (or similar) pugging directly on ceiling.

FLOORS

40.	Joist floor with t. and g. boards and lath-and-plaster ceiling.
41.	Joist floor with floating wood raft finish and 10 mm plasterboard and skim-coat ceiling.

About 35 dB

42.	Joist floor with t. and g. boards and 10 mm plasterboard and skim-coat ceiling.

About 30 dB

43.	Joist floor with t. and g. boards and 10 mm plasterboard ceiling, joints filled and lined with paper.

About 25 dB

44.	Joist floor with plain-edge boards and 10 mm plasterboard ceiling, joints filled and lined with paper.

About 20 dB

45.	Joist floor with t. and g. boards and no ceiling.

Note: A concrete floor is defined as one of not less than 220 Kg/m^2 overall average weight. It can be taken to include floors with hollow clay pot or concrete units or trough-shaped beams, provided that the ribs are integrated into the structure by concrete fillings in the finished floor. It should *not* be taken to include floor slabs freely supported on unintegrated beams unless such slabs themselves weigh at least 220 Kg/m^2 including any topping screed.

TABLE VIII

TABLE OF RELATIVE IMPACT INSULATION OF FLOORS

The impact insulation of floors is mostly dependent on the nature of the floor finish, and to some extent on the rigidity of the connection between the floor surface and the rest of the floor. The greatest improvement in impact insulation is obtained by the use of a really soft floor finish such as thick carpet on felt, but there are many occasions when such a finish is impractical. Then the second means of improvement mentioned above and exemplified in floating floors must be used.

The following table gives a rating of all the fifteen different floor constructions given in the table of air-borne sound insulation above, with three different degrees of finish softness. These are:

(a) Soft finish, meaning thick carpet with or without underfelt; rubber or linoleum with sponge-rubber backing; plastic sheeting or thick linoleum with hair-felt backing; thick linoleum laid on 12 mm soft fibreboard; thick (not less than 10 mm) cork tiles.

(b) Medium soft finish, meaning thin carpet or matting; cork tiles (less than 10 mm); thick linoleum; wood boards; soft plastic or rubber sheeting or tiles.

(c) Hard finish, meaning granolithic, terrazzo or cement screed; clay, concrete or thermoplastic tiles; marble or stone; pitchmastic; magnesium oxychloride; asphalte; thin linoleum; wood blocks.

Floor Finish			Impact Insulation Rating
Soft	Medium Soft	Hard	
31 to 41 inclusive	32, 35		Very good
42, 43, 44	36, 37, 38	32	Good
45	31, 33, 34, 41		Poor
	39, 40, 42, 43, 44, 45	31, 33, 34, 36	Very poor

Numbers refer to Table VII.

MOVABLE WALLS

There is a frequent demand for some form of movable wall, such as a sliding and folding partition with some reasonable degree of sound insulation. It is never possible to obtain very high insulation from these structures because their weight must be low enough to make them manageable and because they inevitably have a number of more or less imperfectly sealed cracks.

The normal type of sliding, folding partition composed of a number of door-like units hinged together should not be expected to have a higher insulation than that of a similar single door, i.e. about 15 dB, and it may be as low as 10 dB. By very careful attention to the details of crack sealing, for example by using effective rubber seals at all edges of all units, it is possible to raise the insulation almost to the mass value for the actual panel units, say 20 dB for 3 mm hardboard on both sides of a hollow-core construction. By improving the panels (which will mean making them heavier) and still retaining the crack-sealing methods, it is possible to raise the insulation to about 25 dB. To get any higher than this with a simple multi-unit partition is probably not possible, but two partitions of the 25-dB type described above and fixed with at least an 200 mm gap between them will yield about 40 dB insulation.

An insulation of about 40 dB can also be obtained by a single heavy partition if it takes the form of one large unit (thereby

SOUND INSULATION AND NOISE CONTROL PRACTICE

reducing the number of cracks) which can be raised or slid side-ways, and which can be really well sealed at all edges by screw clamps on to soft rubber seals when closed. The panel itself must of course be able to provide at least 40 dB insulation, which will need a weight of 120 Kg/m² if it is a single leaf, but could be less than this if a properly designed multi-leaf panel is used.

Sealing of Cracks

As can be seen from the above tables, adequate sealing of the cracks round the opening lights in windows and round doors will produce a definite improvement in their sound insulation. The most common and least effective method of doing this is by applying strips of felt, sponge rubber or plastic to the stops against which the windows or doors close. This system can be effective if the fastener pushes the window or door hard up against the soft strip all round the opening. It is, however, rare to find this requirement fully met, either because of poor work-manship or more often because windows and doors so frequently warp after they are hung.

An alternative and better solution is to use the springy phosphor-bronze draught stripping material or one of the inter-locking metal draught strips. A good method of sealing employ-ing sorbo rubber strip which allows for some degree of door warping is illustrated in Fig. 70, and another one using an extruded rubber strip is shown in Fig. 77.

None of these methods can be used on the lower edge of a door when there is no threshold or sill. For this position one of the drop-bar type draught sealers is effective. If the door opens over a carpet which can extend right under the whole thick-ness of the door when it is closed, then a seal on this edge can be omitted, provided the door is fitted so that it just meets the top of the carpet when closed.

Sound Locks

A sound lock is (by analogy with an air lock) simply a sound-reducing lobby formed between two doors. The larger this lobby can be made, and the more sound-absorbent material that can be put in it, the better the overall insulation will be. The very small lobby formed by hanging both doors in one wall opening, one at each face, is usually not very effective, unless both doors can be fastened with some positive form of latch. If this is not done, and door-closing springs are relied upon, as one door closes the air trapped in the lobby forces the other one open. If this type of sound lock must be used there should be a minimum distance of 200 mm between the doors and all the reveals and preferably

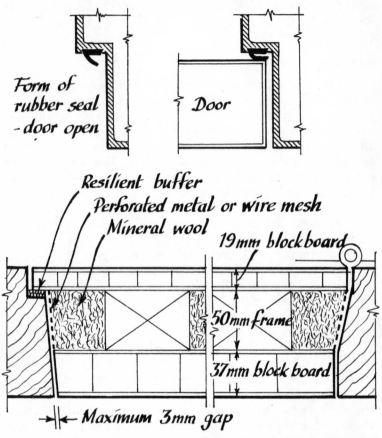

Fig. 77. An Extruded Rubber Door Stop and Seal (above) and a thick Insulating Door with Absorbent-lined Edges (below)

the inner face of one of the doors should be lined with sound-absorbent material.

By far the better type of sound lock is that illustrated in Fig. 78. This is large enough for a person to be able to close one door before opening the other and overcomes all the disadvantages mentioned above. The ceiling and at least one-half of the walls should be sound absorbent, and the walls of the lock must have at least as high an insulation as that of one of the doors.

222

Double Windows

To give increased sound insulation, double windows must have the two glazings as far apart as possible. This will generally mean using two separate frames, although this is not essential. Double glass, with a space of between 3 mm and 12 mm between the panes, such as that which is effectively used for increased heat insulation, will not give more sound insulation than would

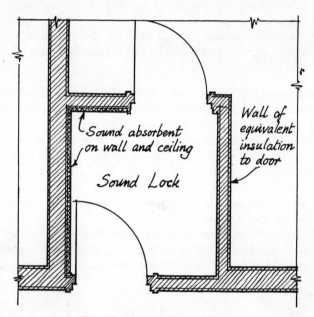

FIG. 78. Plan of a Sound Lock

a single pane of glass of the same total weight. To obtain the improved insulation resulting from the cavity this should never be less than 25 mm and can with advantage be much more. A detail of a double window suitable for studios and having high insulation is given in Fig. 69.

NOISE REDUCTION BY SITE PLANNING

We give here (Table IX) a rough guide to the distance measured along the ground (the oblique distance from the road to windows on higher storeys should be ignored) that should be allowed between roads (kerb) and rooms used for various speech purposes. The conditions assumed are:

- (*a*) the room has windows facing the road,
- (*b*) the road carries continuous heavy traffic, i.e. there are several vehicles passing at one time.
- (*c*) there is no obstruction between the road and the room,
- (*d*) the single windows are of 4 mm glass and tightly closed,
- (*e*) the double windows are of two fully sealed leaves each of 4 mm glass and separated by a 100 mm air-space with absorbent in the reveals,
- (*f*) no person in the room is close to the windows (this last assumption is particularly important for assembly halls and theatres).

If the room has no windows in the side facing the road but has windows in the side walls, then the distance can be halved.

Two distances are given for most of the sets of circumstances. The 'ideal' distance indicates that the intruding noise in the room will be quite negligible. The 'workable' distance indicates that speech communication under the particular circumstances will be possible but with some discomfort; anything less than the workable distance is likely to cause appreciable interference with the use of the room.

When very short distances are given it should be remembered that there is a possibility of noise being made in the room by vibration transmitted through the ground. No general advice on this can be given as it will depend on the nature of the ground.

NOISE REDUCTION BY ABSORBENTS

The value of sound absorbents to reduce noise levels and in some cases enhance effective sound insulation has been explained in Chapter 7. We will comment here on the choice and use of practical materials.

The absorbents available for this purpose are, of course, precisely the same materials as those used for the acoustic correction of rooms and which are described in theoretical terms on p. 59 and listed for performance in Appendix A, p. 309. The absorption coefficient is a measure of the efficiency of a material as a noise reducer, but as this quantity varies so much with frequency, comparison between one material and another may become difficult.

TABLE IX

DISTANCES (IN METRES) BETWEEN NOISE AND ROOM

Room	Window Conditions	Criterion	Distance
Classroom	Open (8 m²)	Ideal	More than 600
		Workable	60
	Single (12·5 m²)	Ideal	45
		Workable	7
	Double (12·5 m²)	Ideal Workable	No restriction
Assembly Hall or Theatre for 500 audience	Open (10 m²)	—	150
	Single (100 m²)	—	30
	Double (100 m²)	—	No restriction
Conference Room for 50	Open (2 m²)	Ideal	300
		Workable	90
	Single (40 m²)	Ideal	60
		Workable	15
	Double (40 m²)	Ideal	15
		Workable	No restriction
Court Room	Open (2 m²)	Ideal	180
		Workable	60
	Single (40 m²)	Ideal	45
		Workable	15
	Double (40 m²)	Ideal	15
		Workable	No restriction
Conference Room for 20	Open (2 m²)	Ideal	225
		Workable	90
	Single (15 m²)	Ideal	38
		Workable	15
	Double (15 m²)	Ideal	9
		Workable	No restriction
Small Private Office	Open (3 m²)	Ideal	225
		Workable	45
	Single (10 m²)	Ideal	15
		Workable	5
	Double (10 m²)	Ideal Workable	No restriction

For the usual kinds of noise, e.g. traffic noise, typewriters and accountancy machines, 'general' factory noise, the simplest and best guide is the average absorption coefficient at the four frequencies 500, 1000, 2000 and 4000 Hz (which for the sake of distinction we have called the loudness reduction coefficient, abbreviated L.R.C.). It will be seen that materials like 5% perforated hardboard, and to a lesser degree other boards of up to 10 or 12% perforation with a porous material behind, have a rather lower rating than some of the other materials such as wood-fibre acoustic tiles.

When a special problem is encountered in which peaks of sound energy occur at some part of the frequency scale outside the range 500–4000 Hz (this would generally best be found by actual noise measurements, including octave analysis, see p. 246) then it is worth considering a type of absorbent treatment which is specially 'tailored' to the requirements, i.e. has highest absorption coefficients where the peaks occur.

The absorbents may be fixed to the ceiling (or underside of the roof) or walls, and the best results will be obtained if they are as close to, and as much surrounding, the noise source as possible. When it is impractical to fix a material to a surface then the absorbents may be suspended freely in the room space, usually above head height. They may either be in the form of 'functional absorbers' (see p. 65) or as simple flat plates (both sides should be absorbent). An example of such a treatment in a top-lit banking hall in which the sound absorbents are integrated into the natural and artificial illumination design is shown in Plate XII.

Sound-absorbent baffle screens are dealt with on p. 236.

VENTILATING PLANT AND OTHER MECHANICAL APPARATUS

The tendency in the design of rotary mechanical equipment is often towards the use of higher speeds and power, accompanied by higher noise levels and more vibration.

The installation of mechanical ventilation plant involves two separate problems. The first is to ensure that the plant itself does not create a noise nuisance in any part of the building, whether it is served by the plant or not. The second is to ensure that the installation of the ventilating ducts does not cause loss of insulation between rooms or between outside noise and the rooms of the building. The simplified diagram Fig. 79 illustrates how these problems are interrelated and shows what effect remedial measures will have on the different aspects of noise control. The measures required will depend on the magnitudes of the noise sources and on the permissible tolerance of noise in the rooms concerned.

The major noise source is usually the fan. Fans are of two main types, axial flow or propeller, and centrifugal. The first type operates at high speeds (1000 r.p.m. upwards) and the second at low speeds (less than 1000 r.p.m.) and their noise characteristics are rather different. The high-speed fans produce

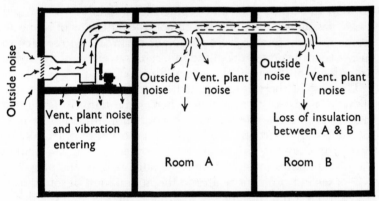

FIG. 79. Siting of a Ventilation Plant

more noise energy in the high-frequency part of the sound spectrum than low-speed fans. The total noise output of both types increases approximately logarithmically with the power, the speed and the size of the fan; that is, doubling the power, speed or size results in roughly a fixed increase in the total noise energy. For detailed information on this subject see Chapter 9, p. 280.

Fans are normally driven by electric motors. The motor itself rarely generates large volumes of noise, although it may cause noise in the structure due to vibration produced by poor bearings or lack of balance in the rotor. The grades of motor known in the industry as 'silent' or 'super silent' are specially engineered to reduce these defects but it should be remembered that these are merely catalogue terms. The noise caused by vibration can be very insidious, especially when the motor is installed on a floor not in contact with the ground. This noise can be prevented by mounting the motor and, if necessary, the fan, on a *properly designed* isolating support (see p. 284). If the motor is mounted on a floor in contact with the ground there is usually no advantage, from a sound insulation point of view, in using isolating supports.

The other sources of noise commonly found in ventilation plant rooms are air-filtering and washing equipment and circulating pumps. Both of these make much less noise than the fans with which they are associated and can therefore usually be ignored.

It is seen from Fig. 79 that noise produced in the ventilation plant room can be disseminated in two ways. It can pass through the walls, floor and ceiling into adjoining rooms, and it can pass along the ducts which act as conductors of sound to rooms near or remote from the plant room. It is necessary to ensure that the noise introduced into rooms in either of the above ways does not impair their normal functioning.

The first of these problems is best dealt with by zoning the plant in a part of the building where noise and vibration can most easily be tolerated, in other words not immediately next to, over or under rooms requiring a high order of quiet. Whatever rooms adjoin, the insulation provided by the ventilation room boundaries must, of course, be adequate for the purpose.

The second problem relates to the attenuation of sound as it passes along (and in some cases through the sides of) the ducts. Attenuation of sound in plain unlined ducts in straight lengths up to the dimensions normally found in buildings is very small, and can be neglected. Some attenuation occurs at bends especially if they are lined with an absorbent, and at changes in section area. The small changes in section which occur at branches provide such small attenuation that it can be ignored, but the 'change of section' which occurs where the duct enters the room provides a considerable degree of attenuation, the amount depending to a large extent on the acoustic conditions in the room. If the duct opening is regarded as a sound source it is obvious that the noise close to the duct will be more than that at some distance from it, and that at a certain distance the sound will have become predominantly 'reverberant' and no further reduction of sound can be expected at greater distances. This critical distance depends on the size of the outlet (source) and on the acoustic conditions, i.e. the amount of absorption, in the room. Generally for average rooms and outlet sizes the distance will be not more than about 3 m.

It is also possible for the grille or louvres over the ventilation opening to produce noise, especially if the air velocity is high and the grille provides considerable obstruction to the air flow. Provided the air velocity is less than 150 m per minute and the vanes or wires of the grille are not seriously obstructive, the noise generated will be too low to take into account.

Further increases in the loss of sound passing along ducts can be obtained by lining the ducts with a sound-absorbent material. These linings should possess the following properties:

(*a*) high absorption coefficient,
(*b*) smooth surface for low air friction,
(*c*) adequate strength to resist disintegration due to air stream,
(*d*) odourless, fire, rot and vermin proof.

When it is impractical to obtain enough sound attenuation by simple lined ducts owing to length limitations, the efficiency can be increased by the use of splitters. These are divisions running along the length of the duct so that it is cut up into a number of channels. Splitters in effect increase the perimeter of the duct without altering its area. They consist of impervious dividing plates which must be faced on both sides with sound-absorbent material.

Another method of getting high sound attenuation in a duct is by using acoustic filters. A filter consists of a section of duct having different dimensions from those of the normal run in which it is placed, and reduces sound in the same way as the grille at the end of a duct. The fortuitous cavities which tend to occur in modern buildings can often be used for ventilation purposes with advantage. The irregular shapes and sizes of the spaces often form excellent acoustic filters if some sound-absorbent lining is present. For example, the space above a suspended ceiling can be used as a 'plenum' chamber from which air can be distributed to the rooms above and below.

Lastly, it is essential to ensure that noise is not introduced into the building via the ducts and that insulation loss is not incurred through their presence. The air inlet and exhaust openings should be kept as far away from intense outside noise fields as possible. For example, an opening on the street-side face of a building would be subjected to a much higher noise level than if it were placed at the top of a short stack on the roof as far away from the street as possible and in the 'acoustic shadow' thrown by the building.

When ducts are constructed of light sheets of metal or other material, there is the possibility of sound passing through the walls of the ducts and thereby 'short-circuiting' some more highly insulating path. For example, if a theatre has a ventilation plant in which precautions have been taken to ensure that neither street noise nor fan noise will pass along the ducts into the auditorium, and a length of lightweight ducting is exposed on the roof, then the possibility exists of street noise passing directly through the sides of the duct and so reaching the

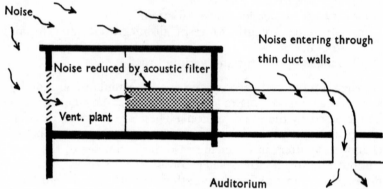

FIG. 80. 'Short Circuiting' of Sound Insulation Measures through Thin Duct Walls

auditorium, having by-passed the acoustic filters or linings which have been incorporated in the system to keep noise out (see Fig. 80).

Another example is where internal bathrooms are used in blocks of flats and they are ventilated by ducts. These will allow noise to get readily from one bathroom to another unless the duct run between the two rooms is very long or unless the duct is lined with absorbent for some distance. It is probable that a total length of lining of 1·5 to 2 m would be sufficient. This can be split into two halves and built into the bathrooms themselves, as illustrated in Fig. 81.

Similarly, ducts which pass through walls or floors can be responsible for lowering the overall insulation whether the adjoining rooms are served by the duct or not. If there are no openings into the duct in the two rooms in question, then the effect on the insulation between the rooms is roughly the same as that of cutting a hole in the dividing wall or floor of the same size as the duct and filling it in with a double thickness of the sheet material of which the duct is made.

Noise sources in heating chambers are very diverse in character, depending on the type of boilers and equipment installed. Noise from mechanically or hand-fired solid-fuel boilers is usually confined to the clatter and clang of the fuel-manipulating tools, or machines, and the defence is to be sought in adequate insulation in the surrounding structure. Oil-fired boilers, particularly of the larger capacities, produce considerable low-frequency noise from the flames in the boiler and are

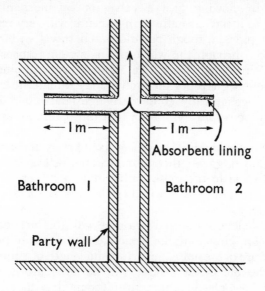

FIG. 81. Sound-baffled Ducts for Bathrooms in Flats or Hotels

commonly associated with fans for draught purposes. It is essential to juxtapose only rooms in which fairly high noise levels are tolerable to such boiler houses, unless special measures to secure very good insulation are taken and especially to provide considerable weight in the structure to insulate the low-frequency components of the noise.

NOISE FROM PLUMBING SYSTEMS AND CONDUCTION BY PIPES

Particularly in buildings where only very low noise levels can be tolerated, e.g. house and flats, it is important to guard against excessive noise from the various plumbing and heating systems. It will be as well first to define what rôle pipes play in the conduction of sound. Where pipes pass through partitions or walls, sound originating in the air outside the pipe is *not* conducted along it to any degree which matters, provided the pipe is a tight fit in the hole through which it passes. When pipes have been suspected of conducting sound in this way it is nearly always found that the sound penetrates through an excessively large hole which has been made in the partition for the pipe. If this sound is generated by the water inside the pipe, for example by flowing past a valve, or by mechanical shock delivered to it such as hammering, then it will very readily pass along the pipes, through partitions and travel with very little attenuation through the whole piping system, especially if the pipes are 'insulated' from their supports by wrapping them in felt. Although the pipes themselves may radiate little sound, if they are mechanically coupled to light partitions or ceilings these may be set into vibration and thus cause an appreciable noise.

On the other hand, if the pipe is fixed firmly at short intervals to the supporting wall, then sound will not be transmitted so far. The wall close to the noise source may, however, be made to radiate noise, although it will do so less if it is a heavy one than a light one.

A flexible length inserted in a pipe will stop most of the noise transmission. How long the flexible length needs to be depends on the nature of the noise, but a 150 mm length of rubber hose or a short flexible metal bellows will be adequate in nearly all cases. The new plastic pipes transmit sound less than metal ones.

Thus the best solution to any problem will depend on the circumstances. For example, if noise close to the source does not matter, then it will be better to fix the pipe firmly to the wall, unless radiation of sound from the wall must be avoided, as in a party wall. If, on the other hand, as little noise as possible is necessary in all rooms, then the pipe should be insulated from all supports and all partitions; flexible lengths should be used close to the noise source, and the whole pipe work cased in substantial ducts.

Probably the most offending device in the matter of noise production is the common ball valve. The usual standard design (although some makes are not so bad as others) creates turbulence in the water system to such an extent that cavitation occurs. This means that numerous small cavities (not air bubbles but nearly perfect vacua) are formed in the water. The shock waves which result when these cavities collapse produce noise, and often erosion of the metal parts of the valve to such an extent that these have frequently to be replaced after quite a short time. The only solution is to use a quieter type of valve and these are now becoming available.

Another source of noise is water and steam hammer, which is caused also by sudden changes in pressure. Pressure changes can be caused by taps (particularly press-operated taps), automatic steam traps or by fortuitous conditions in the piping. If every precaution has been taken to secure quiet operating mechanisms, any further difficulty can only be solved by planning or screening the noisy pipes from the rooms by building them into suitable ducts. Any such ducts must have walls capable of reducing the sound to the desired level, and must be carefully designed, particularly in the matter of access panels, so as to ensure that they do not conduct other air-borne sounds from one part of the building to another.

The noises created by waste water flowing in pipes, although never very loud, can produce annoyance and embarrassment. A simple covering or duct kept out of contact with the pipe and with walls having a weight of 10 to 24 Kg/m^2 will usually reduce such noises to negligible loudness.

Lift machinery was once considered a dangerous source of noise but most manufacturers have so greatly improved their products in this respect that it is no longer a serious nuisance. The only exception might be where manually operated gates are employed; these can be misused so as to create considerable noise.

Noise from Generating Plant, Compressors, etc.

Diesel generating plant and compressors for refrigeration, air supply or vacuum supply are virulent noise sources. Such machinery will usually be sited in a basement room, although cases have been known where it is on a higher floor. The amount of insulation necessary depends on the noise level produced by the machines and that which can be tolerated in the room in question. Planning the noise away from areas requiring quiet is the best solution. Even so it will usually be necessary to ensure that the structure surrounding the machines is substantial—never less than 220 mm brickwork or the equivalent weight in a floor—and that no loopholes are left for the escape of sound.

Anti-vibration Mountings for Noise Reduction

Resilient mountings may be used to solve two problems. One is the isolation of vibrations to prevent discomfort to the tactile senses. The other is to reduce the noise caused by a vibrating object or conversely to reduce the noise in a building or room caused by the vibration of the ground or the rest of the building. These problems are sometimes linked together and mountings which will solve both at the same time are practicable. However, we shall not deal in detail with the first of these problems but concentrate on the second.

The possible frequencies of vibration extend well below the lowest audible frequency (about 20 Hz), but although a vibration may be at a sub-audible frequency it can nevertheless cause noise, either because of harmonics of the vibration at higher (audible) frequencies, or because some part of the building is set into vibration at an audible frequency. Now the isolation of frequencies in the audible range is a comparatively simple matter. It is explained on p. 285 that the static deflection required of a resilient mounting is related to the lowest frequency it is wanted to isolate. For example, a 25-dB reduction in vibration at 25 Hz will be given by a mounting having a static deflection of about 10 mm. But it may be required to isolate a very low sub-audible frequency because of the noise it may cause. To obtain a 25-dB reduction in a vibration of 10 Hz (a typical frequency for vibration caused by electric trains) the static deflection of the mounting would need to be at least 63 mm. However, the insulation of higher (audible) frequencies provided by such a mounting would be very much higher than 25 dB and therefore such a large static deflection might not be necessary, although insufficient experience is available to be certain of this. It would appear to be impractical to isolate complete buildings from these very low-frequency vibrations, but it may well be possible to obtain effective noise isolation of individual rooms in a building.

The amount of static deflection required gives a rough guide to the best type of mounting to use, as follows:

Mounting	Machine Speeds (r.p.m.)	Useful Deflection Range
Steel or rubber springs	up to 700	25 mm upwards
Rubber in shear	700–1200	2 to 50 mm
Rubber, cork or proprietary compounds in compression	over 1200	6 mm and less

Designing mountings is an engineering problem which is commonly undertaken by the specialist firms who supply proprietary mountings and should be competent to do so. Ill-advised (and unsuccessful) attempts are often made to provide isolation by inserting a cork or felt pad under a vibrating machine without any proper consideration of the principles involved.

NOISE IN INDUSTRY

In the design of industrial buildings very many factors have to be taken into account and the need to compromise with the ideal solution to any one problem becomes correspondingly great. The demand for reasonably quiet conditions for workers is growing, and the study of the basic rules at the beginning of this chapter is recommended. It is certain that many industrial processes which give rise to excessive noise will continue to be used, and although every effort should be made to discover means of reducing this at source, for example by improved design of machines or by substitution of some quiet process such as welding for a noisy one such as riveting, there will often be a residual problem to deal with in the building.

There are three aspects of the noise problem. First there is the requirement to ensure reasonable conditions for the individual workman who is making the noise either manually or by his machine. The second is the protection of other workers either within the immediate vicinity of the noise source or at some distance and possibly in some other room. The third aspect is the protection of surrounding property from noise created in the factory.

Town-planning and zoning regulations go some way to reducing difficulties in the last problem. Reference to Chapter 10, p. 307, where some tentative noise criteria are given, may help in deciding on the minimum structure required for a building housing an intense noise source. In single-storey buildings it is very often the roof which is the weakest insulation link. Open windows, ventilators and exhausts also account for a very large number of the known complaints. Some information on sound reduction of ventilation outlets is given in Chapter 9, p. 279.

The protection of workers from general noise, not necessarily of their own making, can be tackled in two ways. The general introduction of sound-absorbent materials can nearly always bring about small but significant reductions in the overall sound levels. In these problems it is often well worth obtaining a frequency analysis of the noise so as to ensure that the absorbents to be introduced have the highest possible absorption at frequencies where it is most needed. The other way is to introduce partial or complete screening either of the man who needs quiet or of the machine making the noise. A well-known example of this principle is the acoustic telephone booth. Such semi-enclosures, lined with sound absorbents, will provide a local area of comparative quiet, but it must be remembered that their effectiveness depends to some extent on the closure of the open side by the body of the person phoning, and this sets a limitation on the practical size. Where rooms (such as offices) are built in a noisy factory area it is very common to see them provided only with screen walls (often glazed) and no ceiling. This type of construction cannot give any significant reduction in the noise, whereas if even the lightest type of ceiling is provided a sound reduction of perhaps 20 dB becomes possible. Even greater amounts of reduction can be had by using heavier ceiling and wall construction, although careful attention to door sealing, ventilation ducts and glazing will then be needed. The alternative approach is to use screens completely or partially round the noisy machines. If the screens are partial (and they should, of course, be made as complete as is practicable) then the amount of sound reduction will not be very great. It may, however, be enough in a given position, that is where the screen provides acoustic shadow, to be well worth while. The screens should always be lined on the inside, and also with advantage on the outside, with a sound-absorbent treatment. Since the overall 'insulation' is unlikely to be more than a few decibels, quite light materials can be used for the screens. For example, if the absorbent faces consist of perforated hardboard over a layer of mineral wool on both faces, a single sheet of unperforated hardboard in the centre of the screen thickness, dividing the two layers of mineral wool, provides an adequate barrier to sound.

When the screen can be made absolutely complete then the amount of sound reduction obtained will depend on the average overall insulation of the complete enclosure. In all cases it is essential to introduce as much sound-absorbent material inside the enclosure as possible. This will limit the build-up of reverberant sound which might otherwise reach a higher value than that before the enclosure was formed, although most of the normal sound absorbents *cannot* be relied upon to add much to the insulation of the walls or ceiling to which they are fixed.

Lastly we must consider the case of the workman making noise. It has often been said that one is not worried by the noise oneself makes, and this is very largely true. Machine-tenders very often need to hear their machines properly as they can detect faulty operation by the noise made. However, the noise of surrounding machines or workmen can be a distraction and there is evidence that the installation of general absorbent treatment either alone or in conjunction with screens in many cases provides a distinct improvement to working conditions.

There still remains the problem of operations which cannot be successfully quietened to harmless levels by any of the means described. (The maximum noise levels tolerable are given on p. 287.) For these the only solution is for the workers to use personal protection in the form of ear-defenders.

Ear Protection

We will not deal here with protection against the extremely high noise levels met with in, for example, jet-engine testing, but only with protection against noise levels of up to about 110 dB which are fairly common in factories. The three types of ear-defenders are:

(a) insert types, i.e. plugs which are inserted into the ear canal and made of rubber or plastic and sometimes with metal weights to increase the mass or with valves that close only at high noise levels,

(b) cushion types, i.e. enclosures made of rubber or plastic which enclose the whole ear, being kept in place with headbands,

(c) helmet types in which the whole head is enclosed.

The helmet type is not normally used in factories, and the cushion type is rare although they are thought by some people to be more comfortable than the insert types. The insert types are the most commonly used but as the shape of the ear canal varies considerably between individuals, the correct size must be chosen. They must be tightly fitting because quite small leaks considerably reduce the effectiveness.

The order of protection afforded by insert types and by cushion types is shown in Fig. 82. If an octave analysis of the noise in a factory is available, the attenuation provided by

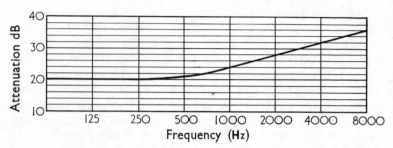

Fig. 82. Protection provided by Cushion or Insert Ear-defenders

such ear-defenders can be subtracted from the noise levels and the results compared with the deafness criteria (p. 287). In most factories this protection will be adequate, but in very noisy cases, insert types plus cushion types can be used.

.9

Sound Measurement and Calculation

In the first part of this chapter we describe briefly the apparatus used for making noise measurements, excluding the more complicated equipment which will only be employed by those who specialise in the subject. In the second part we describe the types of noise measurements that can be made, and give the various relationships and corrections that are useful with such measurements. Many of these relationships and corrections are only approximate, either because the conditions are such that precise calculations are not possible or because there is, at the present time, not sufficient practical knowledge.

In the succeeding chapter we give the various criteria—again many of them are only approximate—that can be used to assess the subjective effect on people of known noise levels.

The accuracy with which noise levels can be measured depends on the circumstances, but it is seldom that there is any need to give the results more accurately than to the nearest decibel. To be on the safe side, noise level measurements should be rounded-off upwards and sound insulation measurements downwards.

APPARATUS

MICROPHONES

The first part of any measuring equipment is the microphone, and three types are commonly used for field measurements. The first type is the moving-coil microphone. This consists of a diaphragm to which is fixed a light coil of wire. As the diaphragm moves under the action of the alternating sound pressure, the coil moves in the field of a magnet. The movement generates a voltage in the coil which is a 'copy' of the sound pressure. Moving-coil microphones have a low electrical impedence—

Engineering

Limerick Technical College

usually 20 to 40 ohms—and a long cable may therefore be used between the microphone and the amplifier. This makes them particularly suitable for some types of field measurements; although all microphones are fragile, moving-coil microphones are the least fragile.

The second type is the condenser microphone. This consists of a light diaphragm spaced a small distance from a back plate. A d.c. voltage is applied between the diaphragm and the plate: the sound pressure moves the diaphragm relative to the plate, thus altering the capacitance. The alternating voltage thus produced is fed through a condenser to isolate the d.c. voltage— to the first stage of the amplifier. Condenser microphones have

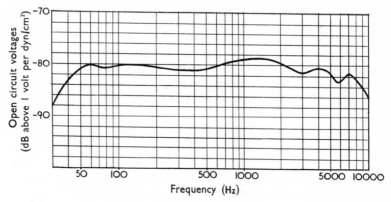

Fig. 83. Moving Coil Microphone Calibration Graph

a high electrical impendence, and thus the first stage of amplification must be close to the microphone.

The third type is the crystal microphone. This is not so commonly used: it employs a crystal with piezo-electric properties, and like the condenser microphone, has a high impedance.

Any microphone must be absolutely calibrated and this is usually done by the manufacturer. With some types of apparatus (for example, a sound-level meter) the microphone is 'built-in', and the manufacturer should have calibrated it so that the meter readings will be correct. On this type of instrument there is often a pre-set gain control so that the gain of the amplifier can be kept constant.

The sensitivity of microphones may vary with frequency: the less the variation the better. A typical calibration graph for a good moving-coil microphone is shown in Fig. 83. Good quality condenser microphones, however, can be built to have practically no variation over the audio-frequency range.

When a microphone is not 'built-in', the absolute calibration of the microphone is required in terms of the voltage produced by the microphone when placed in a field of known sound pressure, as a function of frequency. The voltage given in the calibration is usually that produced by the microphone when on open-circuit. With low-impedance microphones this will be affected by the impedance of the circuit it works into.

Unlike microphones used for speech-reinforcement systems, which are usually deliberately made directional, microphones used for noise measurements are non-directional, except for very special purposes. However, because of their size these microphones unavoidably become directional at the higher frequencies (above 1000 Hz). The absolute calibration must therefore take account of whether the calibration sound field was incident from one direction only, or from all directions (random sensitivity), and the appropriate calibration must be used depending on the type of measurement.

It cannot be too strongly emphasised that all microphones are fragile, and if damaged, often do not stop working altogether but give out a smaller voltage (never greater) than their calibration. It is thus very easy to get noise readings much lower than they should be, unless the microphone sensitivity is checked regularly. It should be noted that if a moving coil microphone becomes defective in such a way that it generates less volts than it should while still keeping the same internal impedance, the electrical calibration (described below) will be unchanged and therefore the fault may go undetected. As a check against such faults, a standard noise source can be used. This may consist of some small device, e.g. steel balls falling on to a plate, put in a standard position relative to the microphone and which can be relied on to produce the same noise each time it is operated. Even this is not a complete check, because such devices do not usually produce much low-frequency noise, and a common fault in moving-coil microphones (when they are dropped) is the loss of the low-frequency response only. Thus when used to measure low-frequency noises they would give wrong answers, but would still respond correctly to the higher-frequency output of the check noise. At least this is so when no form of frequency analysis is being made; when such an analysis is made using the check noise, this 'low frequency only' fault would be shown up.

The lower limit of noise that microphones will measure down

to is set by the unavoidable electrical 'noise' of the circuit. This lower limit varies with frequency and with the type of microphone, for moving-coil microphones being usually below the threshold of hearing at low frequencies, but above it at medium and high frequencies. In other words, low-frequency noises can usually be measured by moving-coil microphones even when they are just at the limit of audibility, but at other frequencies the ear can detect sounds which the microphone will not respond to.

When microphones are used out-of-doors it is nearly always necessary to protect them from the wind, which otherwise will make a lot of noise due to turbulence round the microphone. This is done by using silk or muslin or nylon stretched over a wire frame completely enclosing the microphone. The volume enclosed should be as large as possible, and for very severe conditions it may be necessary to enclose the first wind-shield by a second one.

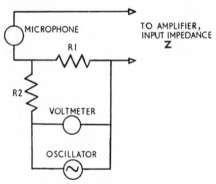

Fig. 84. Electrical Calibration Circuit

ELECTRICAL CALIBRATION

All sound pressures are small, and the corresponding microphone voltages are small. Amplification is always necessary, and in addition to the absolute calibration of the microphone the electrical calibration of the apparatus is required. For moving coil microphones this is usually done by the circuit shown in Fig. 84. R_1 is small and R_2 is large, say 1 ohm and 10,000 ohms, and both values must be accurate. A known voltage from the oscillator is applied, and the corresponding current through R_1 produces a voltage across it and thus across the microphone and the input impedance Z, in series. The voltage across R_1 can thus be regarded as equivalent to the open-circuit voltage of the microphone. An absolute calibration of a microphone in terms of its open-circuit voltage can thus be compared directly with the voltage across R_1.

For example, the open-circuit absolute calibration of a microphone is usually given as so many dB above 1 volt per dyn/cm². Thus if, at a certain frequency, this sensitivity is given as – 80 dB, this means that when this microphone is in a sound field of pressure 1 dyn/cm² it will give an open-circuit voltage of 80 dB less than 1 volt, or 20 dB less than 1 millivolt. Now 1 dyn/cm² is 74 dB above the usual reference pressure of 0·0002 dyn/cm², therefore this particular microphone at this particular frequency would give an open-circuit voltage of 94 dB below 1 millivolt (– 20 – 74 dB) for a pressure of 0·0002 dyn/cm². 10 volts from the oscillator (Fig. 84) will produce 1 millivolt across R_1, and the measuring amplifier will give a reading of, say, x dB. Then we know that a reading of x dB indicates a pressure level of 94 dB above 0·0002 dyn/cm² at this frequency.

Measuring Amplifiers

The range of sound pressures met with in practice (see p. 29) is from 0·0003 dyn/cm² up to 300 dyn/cm², i.e. a factor of one million, so even if the lower limit (set by the electrical noise) is say, 0·03 dyn/cm², a measuring amplifier still has to deal with voltages in the ratio of 1 to 10,000. Microphones themselves are designed to respond linearly over this range of pressures (i.e. if the pressure increases by a factor of two, so does the voltage output) and in fact most common microphones respond linearly up to about 140 dB. Above that, they will tend to give distorted results, and may be permanently damaged. However, the amplifier following a microphone will not usually be able to handle such a big voltage range, and obviously a meter scale could not show it. It is usual to put calibrated attenuators in the amplifier circuit which are adjusted by hand for the particular noise being measured so that the meter reading is convenient. The meter scale is calibrated in decibels, so the total noise level will be given by the sum of the attenuator setting and the meter reading.

The detailed design of measuring amplifiers is beyond the scope of this book.

Indicating Devices

After the microphone voltages have been amplified they must be applied to some indicating device and the simplest is a meter.

The audio-frequency voltages must be rectified in some way before they will operate a meter. It is better, for various reasons, if the meter circuit is designed to measure the root-mean-square value, but often the mean rectified level (or sometimes the peak level) is measured. The time constants of the meter are not usually important and either the meter itself can be fairly slow in operating or a condenser can be used in the circuit so that the time constant is about 0·1 second. If the noise level is fluctuating rapidly and the average value is required, then it is sometimes helpful to have a heavily damped meter which slows up these fluctuations. This makes the meter easier to read, but of course the noise is actually fluctuating and, depending on the nature of the measurements, it may be desirable to read also the maximum and minimum values.

If the noise is transient only, e.g. a train passing, then while the maximum value can probably be read off a meter without too much difficulty it will not be possible to get any measurement of how the noise level varies with time. For these cases some form of level recording instrument is desirable. An ordinary pen recording-milliameter could be used in place of the meter but this will have a linear scale. That is to say, if its full-scale deflection is, say, 10, and the minimum deflection that can be read is 0·5, this factor of 20 is only 26 dB, which may not be a big enough range. What is more, the answers will be wanted in dB above 0·0002 dyn/cm^2, and all the linear readings on the pen recorder will have to be converted into logarithmic units. A much more useful (but more elaborate) recorder to use is what is called a 'high-speed level recorder'. This type of recorder traces out on the recording paper (usually a stylus scratching on waxed paper 5 cm wide) the logarithm of the voltage applied to the instrument. Thus it can be calibrated—in connection with the microphone and amplifier—to read directly in decibels. Its range can be selected by using interchangeable input potentiometers, the most commonly used range being 50 dB but 25 dB or 75 dB ranges are also available. The speed at which the paper travels through the recorder can also be selected at will, from a few centimetres an hour (for use when noise levels over a long period are required) up to 100 cm a second (when noises which are changing in level very rapidly are being measured). This type of recorder will respond very quickly if desired—at a rate of up to about 1000 dB per second—but its response rate can be slowed down as required.

The best level recorder will measure either the root-mean-square value of the noise level, the average value or the peak value, as required. We are now of course, referring to the values of the wave-form—not the average and peak (or maximum) values of the total noise.

Occasionally, the exact wave-form of a noise is needed, and then the output from the amplifier is connected to a cathode-ray oscilloscope and the displayed wave-form observed or photographed. This technique is limited usually to the investigation of the sources of machine noise where it may be desirable to relate the wave-form with a particular motion of the machine. It should be noted that for this type of analysis (unlike the other noise measurements described below) the phase changes caused by the microphone and measuring apparatus are important.

ANALYSIS

Of much more general application is some form of frequency analysis. This means that electrical filters are introduced into the measuring equipment so that only the noise in one frequency band reaches the meter or level recorder at one time. Such band-pass filters are of two basic types. The first is the constant bandwidth type, and a common bandwidth is 5 Hz (although other values are available). This means that, for example, when the filter is set to 100 Hz only the noise between 97·5 and 102·5 Hz is measured. If it is set to 1000 Hz then the band passed is 997·5 to 1002·5 Hz. It is obvious that this type of filter gives a very detailed analysis of a noise, often in fact too detailed for practical use.

The other class of analyser has a constant percentage bandwidth (i.e. a bandwidth which is a constant percentage of the centre frequency of the band). They are also known as proportional analysers (i.e. the bandwidth is proportional to the centre frequency). These analysers can pass either what are known as narrow bands or broad bands. A common example of a narrow-band analyser is one with a bandwidth of 3%, that is to say it passes frequencies lying within $1\frac{1}{2}\%$ of the frequency it is set at. Thus, for example, at 100 Hz it would pass 98·5 to 101·5 Hz, and at 1000 Hz it would pass 985 to 1015 Hz.

For nearly all the purposes of this book a broad-band constant-percentage analyser is of most interest, and the two sorts we are concerned with are 1/3rd octave filters and 1 octave filters. 1/3rd octave filters pass of course bands of noise one-third of an octave wide, and the centre frequencies of successive bands are thus 1/3rd octave apart. The results of octave or 1/3rd octave analyses are usually presented as a graph with each band represented by a point at the centre frequency. The preferred centre frequencies for 1/3rd octave analyses and the most commonly used octave filters, together with their centre frequencies, are given in Table X.

TABLE X

⅓rd Octave Centre Frequencies (Hz)	Octave Bands (Hz)	Centre Frequencies of Octave Bands (approx.) (Hz)	Preferred International Octave Bands and Centre Frequencies (Hz)	
(Lower than 50 if necessary)			22–44	31·5
50	37–75	53		
63			44–88	63
80				
100	75–150	106		
125			88–175	125
160				
200	150–300	212		
250			175–350	250
315				
400	300–600	425		
500			350–700	500
630				
800	600–1200	850		
1000			700–1400	1000
1250				
1600	1200–2400	1700		
2000			1400–2800	2000
2500				
3150	2400–4800	3400		
4000			2800–5600	4000
5000				
6300	4800–9600	6800		
8000			5600–11,200	8000
10,000				

MAGNETIC TAPE RECORDING

If a frequency analysis of a transient noise is required, then obviously it will be necessary to repeat the measurement for each frequency band. Thus for octave-band analysis eight measurements would have to be made. A more convenient way is to record the noise on magnetic tape and replay it subsequently through the octave bands on to a level recorder. This order of technique is rather beyond the scope of this book, but a few points to be watched will be mentioned. An electrical calibration in terms of the calibration of the microphone should be recorded on the tape at the start and finish of each set of measurements. The gain control of the tape recorder (which usually controls both the recording level and the play-back level) should be calibrated in decibels. Good-quality tapes do not vary in sensitivity along their length by more than ± 1 dB. If the electrical calibration is recorded as a glide tone rather than as discrete frequencies, the response of the filter will be drawn out automatically on the level recorder during replay and will serve as a check on the operation of the circuit. The recording level on the tape should be as high as possible while not overloading it; this is because the signal-to-noise ratio of the tape recorder will probably not be as good as might be desirable for recording a noise which is changing a lot. The tape machine should be reasonably free from 'wow' and 'flutter', but for noise measurements this is not as critical as it is for speech or music. The speed of the machine must be constant.

MEASUREMENTS

For all practical purposes it is the pressure of a sound field that is measured, and all of the various techniques described in this book are related to pressure measurements. However, it should be realised that the pressure does not completely specify a sound field. For example, the relation between pressure and the energy of a sound wave is different for plane waves and for reverberant sound fields. For this reason some of the relationships given in this chapter are not obviously understandable in terms of pressures, and the reader is referred to the standard text-books on physical acoustics if he wishes to understand their derivations.

Noise Measurements Out-of-Doors

(a) Noise Levels at Given Positions

The simplest of all noise measurements is that of the noise level at a given position or positions. An example would be a measurement of the noise levels at positions in a residential area for comparison with the criterion described on p. 303. The microphone is simply put at appropriate positions and the readings taken. The only point to be remembered is the directionality, at the higher frequencies, of the microphone. If the noise is coming mainly from one direction, which it usually will be in the open air, then the microphone calibration relating to sound incident from this particular direction should be used. If the noise is random, i.e. coming from more or less all directions, then the random sensitivity should be used.

(b) Noise Sources

Most noise sources are directional to a greater or lesser extent, and thus when a noise source is being measured out-of-doors it will be necessary to measure at several positions to get a complete picture. Measurements should not be made too close to the source, because this is the so-called 'near-field' region where irregularities will occur. Neither should the measurements be made too far away, because the attenuation introduced by the atmosphere will then be significant. In general, the distance should be at least several times the greatest dimension of the source and at the most about 60 m. There must be no nearby reflecting surfaces such as walls.

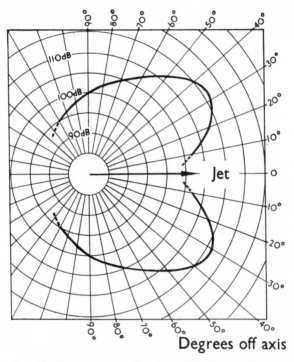

Fig. 85. Sound-pressure Levels in 300–600 Hz Octave at 30 m from Jet Engine

An example is shown in Fig. 85. This gives the sound pressure levels in the 300 to 600 Hz octave measured at a distance of 30 m from the source, a jet-engine running in the open air.

(c) Effect of Distance

The first factor affecting the reduction of noise level with distance from the source is the inverse square law, which is that, due to the spherical spread of the sound, the pressure is inversely proportional to the distance. Although this is only strictly true when there are no reflecting surfaces nearby, in practice it can be used out-of-doors even although there is the ground surface present. Thus the practical procedure is, from the measurements at a given distance, to allow for the inverse square law (using

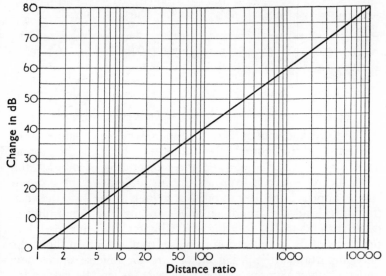

FIG. 86. Reduction due to Inverse Square Law

the pressure ratios of dB's given in Appendix E on p. 321, e.g. a change of 10 dB for a factor of 3·2 in distance, or Fig. 86); the additional factors listed below are then added as appropriate.

The second factor is the molecular absorption of sound by air. This varies with temperature and relative humidity, but is only significant at frequencies of 1000 Hz and upwards. The exact values have been investigated in several laboratories, but the results are not very consistent either between themselves or as compared with the few measurements that have been made in the open air. As a rough working rule, the figures given in Table XI can be used.

TABLE XI

MOLECULAR ABSORPTION OF SOUND BY AIR

Octave Band (Hz)	Attenuation in dB/1000 m					
	21°C			2°C		
	Relative Humidity			Relative Humidity		
	40%	60%	80%	40%	60%	80%
600–1200	3	3	3	10	6	0
1200–2400	13	6	6	33	16	3
2400–4800	33	16	16	49	49	33
4800–9600	130	82	49	82	130	82

The third and fourth factors, namely wind and temperature gradients and the ground attenuation (i.e. the effect the ground has on the sound) do not affect the propagation of sound in the vertical direction, e.g. from an aircraft in flight to the ground. Nor, it appears, are they important so long as the path between the source and the ground subtends an angle at the ground of more than 5° to 10° (see Fig. 55, Chapter 7). Thus for many problems of the propagation of noise from aircraft in flight the only important attenuating factors are the inverse square law and the molecular absorption. However, when the subtended angle is small, i.e. when the sound is propagated nearly horizontally, then the gradients and the ground attenuation become most important.

Unfortunately, the propagation under these conditions is such a complex phenomenon that there is so far little practical information available. One of the many complications is that the attenuation will often depend on the absolute distance and cannot be expressed as so many dB per 1000 m.

The effect of a wind gradient is to reduce the intensity of the sound up-wind and perhaps to increase it down-wind. The effect of a temperature gradient during the day-time (when usually the temperature decreases with height) is to decrease the sound level at a distance from the source depending on the magnitude of the temperature gradient and on the height of the source above the ground. Conversely, at night when the temperature increases with height the sound level may be increased at a distance.

Again, there is little practical information about the effect of the ground on horizontal propagation. This ground attenuation is not proportional to distance and, in theory, at large distances has the effect of changing the inverse square law to an inverse fourth power law, i.e. the sound pressure is reduced 12 dB for every doubling of the distance. However, for practical purposes we can only assume that the attenuation is proportional to distance and that when the ground is hard, e.g. concrete, then the ground attenuation is negligible; when the ground is covered with grass we can use the approximate figures given in Table XII.

TABLE XII

GROUND ATTENUATION OVER GRASS

Octave Band Hz	37–75	75–100	150–300	300–600	600–1200	1200–2400
Attenuation (dB/1000 m)	3	10	23	30	23	10

For example, suppose it is necessary to calculate the sound level in the 300–600 Hz octave at a distance of 600 m from the source shown in Fig. 85 and on the 30° axis. From Fig. 85 it is seen that the level on the 30° axis at the measuring distance of 30 m is 116 dB. The reduction due to the inverse square law from the distance ratio of $600/30 = 20$ is (from Fig. 86) 26 dB. If the ground is grass-covered the ground attenuation is, from Table XII, $\frac{600 \times 30}{1000}$ dB $= 18$ dB approximately. The total reduction is thus $26 + 18$ dB $= 44$ dB which subtracted from the original 116 dB level gives a level of 72 dB.

(d) Walls

To reduce the transmission of sound out-of-doors some obstacle such as a special wall or a building can be interposed between the source and the receiving position. The reduction obtained by such a wall is limited by diffraction over or round the wall and by the scattering of the sound by the inhomogeneity of the air. Further, if any appreciable ground attenuation is present it may be lessened (and thus the reduction due to the wall lessened) because the path of the sound over the wall is further away from the ground than the direct path would have

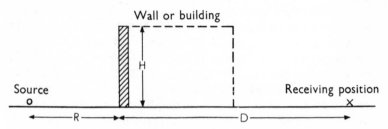

FIG. 87. Wall between Noise Source and a Receiving Position

been. All these factors lead to much uncertainty about the reduction likely to be obtained in practice, but the following method will provide a rough guide.

The wall relative to the source of noise and the receiving position is shown in Fig. 87.

The parameter X is given by:

$$X = \frac{2[R(\sqrt{1+(H/R)^2}-1)+D(\sqrt{1+(H/D)^2}-1)]}{\lambda[1+(H/R)^2]} \quad (7)$$

or, where $D \gg R$ and $R \geqslant H$,

$$X \fallingdotseq \frac{H^2}{\lambda R} \qquad . \quad . \quad . \quad . \quad (8)$$

where λ is the wavelength of the sound in air.

The reduction (dB) in sound level at the receiving position due to the interposition of the wall is given by:

$$\text{Reduction} = 10 \log 20X$$

This relationship is plotted in Fig. 88.

'Flanking' transmission round the ends of the wall can be neglected if the wall is made long enough so that the distance from the source to the ends of the wall is at least twice the distance R. Or the wall can have 'wings' brought back so as to enclose the source on both sides.

The correction for the loss of the ground attenuation can only be extremely approximate, but it may be taken that the ground attenuation is halved by the presence of the wall.

Wind and temperature gradients cannot be considered because of the lack of practical information, but their effect on the reduction due to a wall is probably small. However, some correction can be made for scattering of the sound over the wall

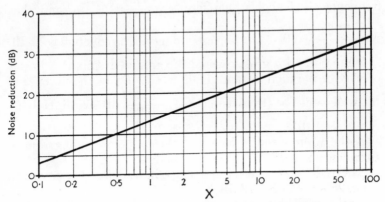

Fig. 88. Relation between Noise Reduction and X

due to turbulence of the air above the wall. The correction figures, i.e. the amount by which the reduction will be lessened, are given in Table XIII, for various wind speeds (because turbulence is related to wind speed).

For example, suppose we wish to calculate the reduction due to a wall 6 m high interposed between a source situated 9 m

TABLE XIII

LESSENING OF REDUCTION DUE TO TURBULENCE (IN dB)

Octave Band (Hz)	Wind Speed		
	8 k.p.h.	16 k.p.h.	32 k.p.h.
37–75	0	0	1
75–150	0	0	3
150–300	0	1	6
300–600	0	2	8
600–1200	0	4	10
1200–2400	1	7	13
2400–4800	3	9	16
4800–9600	8	14	20

from the wall and a point 90 m on the other side of the wall, i.e. $H=6$, $R=9$ and $D=90$ (Fig. 87). Then X (from $X \doteqdot \dfrac{H^2}{\lambda R}$) is calculated for each octave band, each band being represented by its mean frequency. For the 600 to 1200 Hz band the mean frequency is 840 Hz therefore $\lambda = \dfrac{335}{840} = 0\cdot4$ m and thus $X = \dfrac{36}{0\cdot4 \times 9} = 10$. The corresponding noise reduction is 23 dB.

If the ground is grass-covered the ground attenuation in this octave for the total distance of 99 m $(R+D)$ in the absence of the wall would be (from Table XII) $\dfrac{23 \times 99}{1000}$ dB $\doteqdot 2$ dB. Halving this gives 1 dB.

If the wind speed is 16 k.p.h. the loss of shielding due to turbulence is 4 dB.

The reduction in level actually obtained for this octave is, then, $23 - 1 - 4$ dB $= 18$ dB.

Similar calculations for the other octave are displayed in Table XIV. Actually, once X and the noise reduction have been calculated for any octave the noise reduction for the other octaves can be obtained simply by adding 3 dB for every octave increase.

TABLE XIV

CALCULATION OF SOUND LEVEL DUE TO WALL

Octave Band (Hz)	Mean Frequency (Hz)	X	Noise Reduction (dB)	Corrections (dB)		Final Reduction (dB)
				Ground Attenuation	Turbulence	
37–75	53	0·7	11	0	0	11
75–150	106	1·3	14	0	0	14
150–300	212	2·7	17	1	1	15
300–600	425	5·1	20	1	2	17
600–1200	850	10	23	1	4	18
1200–2400	1700	21	26	0	7	19
2400–4800	3400	42	29	—	9	20
4800–9600	6800	85	32	—	14	18

(e) Trees

If the sound travels any appreciable distance through belts of trees then some additional attenuation will be introduced. Table XV gives some approximate attenuation figures (after Eyring) for woods or forests in leaf where the density of the trees and undergrowth is such that it is just possible to see a moving white object at 61 m.

TABLE XV

Octave Band (Hz)	Attenuation (dB/100 m)
37–75	2
75–150	3
150–300	5
300–600	6
600–1200	7
1200–2400	10
2400–4800	16
4800–9600	23

NOISE MEASUREMENTS INDOORS

(a) Noise Levels at Given Positions

The simplest measurement indoors (as for out-of-doors) is that of the noise level at a given position or positions. An example is when it is required to know the noise levels an operator of a machine is subjected to. The microphone is put where his head would be, and the measurements made. If the noise is coming mainly from one direction (as it probably would be in this example) then the microphone calibration for sound incident from this direction should be used. If the sound is coming from more-or-less all directions (as it probably would be in a room at positions not close to the noise source), then the random sensitivity should be used.

(b) Direct and Reverberant Sound

As explained in Chapter 2, when a noise source is operating in a room the direct sound will predominate at positions close to the source and the reverberant sound at positions away from the source. If the source is non-directional then the direct sound level will equal the reverberant sound level at a distance, r, in metres, from the source given approximately by:

$$r = \frac{\sqrt{A}}{7} \qquad . \quad . \quad . \quad . \quad . \quad (9)$$

where A is the total sound absorbent present in the room, in square metre units.

If, as is more usual, the source is directional the conditions are much more complicated and will not be dealt with here. The only uncertainty caused by not dealing with them is the determination of the distance at which the direct sound ceases to predominate. However, in practice it is nearly always possible to decide merely by listening which type of sound is predominant.

Unless it is a question of determining the levels at given positions, as described above, it will usually be the reverberant sound which is of more interest. This is because the reverberant sound level is required either (*a*) for estimating the reduction in this sound level that would result from the introduction of extra sound absorption into the room, or (*b*) for most calculations concerning sound insulation.

The reverberant sound level in a room will not be uniform unless the room is so large and irregularly shaped that the sound field is completely diffuse, and this is seldom the practical case. The level should be measured at several positions to obtain an average figure. For various reasons it is better to take the arithmetic mean of the absolute pressures or the absolute intensities, but as in practice the individual readings will be in decibels it is simpler to average these readings. If the spread of the decibel readings is of the order of 10 dB it is better to average these values and then to add 1 dB.

The measuring positions should be chosen so that it is the reverberant sound which is being measured (and, as we have said, this can usually be decided by ear) but they should not be close to any of the room surfaces. This is because, depending on the nature of the surface, the pressure close to the surface (i.e. less than half a wavelength away) will be up to 6 dB higher than the reverberant sound pressure.

(*c*) *Absorption in Rooms*

The absorption coefficient of a material indicates the proportion of sound absorbed by the material out of the total sound incident on it. For example, an absorption coefficient of 0·3 (or sometimes expressed as a percentage, i.e. 30% in this example) indicates that 30% of the sound incident on the surface of the material will be absorbed by it, the other 70% being reflected. The total absorption of a surface is given by the absorption coefficient multiplied by the area. Thus if the area of a surface of absorption coefficient 0·3 is 20 m², then the total absorption is 6 sq. metre units, or sabins.

The absorption coefficients of materials vary with frequency. Appendix A on p. 309 gives the values for several materials. These values are given at the specific frequencies used in room acoustics, but it will be sufficiently accurate for noise problems to take the frequency from Appendix A nearest to the centre frequency of any particular octave band. For example the centre frequency of the 300 to 600 Hz octave is 425 Hz (see p. 246) and the nearest frequency in Appendix A is 500 Hz. Materials in Appendix A for which only the values at three frequencies are given must be dealt with by extrapolation and interpolation. However, extrapolation to the highest octave can not be done, but this will not usually be important because of the air absorption (see below).

The absorption due to people and seats is usually given as so many square metre units per person or seat; these values are also shown in Appendix A.

At frequencies of 1000 Hz and upwards the absorption of the air becomes important, and this depends on the temperature and the humidity. The approximate values of air absorption for a temperature of 20°C and for various relative humidities are given in Table XVI (based on the measurements of Evans and Bazley). The absorption in sabins is obtained from $x \times V$ where V is the volume of the room in cubic metres.

TABLE XVI

ABSORPTION OF SOUND IN AIR AT 20°C (x PER CUBIC METRE OF VOLUME)

Octave Band (Hz)	Relative Humidity		
	40%	60%	80%
600–1200	0·0016	0·0016	0·0016
1200–2400	0·007	0·007	0·005
2400–4800	0·026	0·02	0·016
4800–9600	0·07	0·052	0·033

The total absorption of a room can be calculated by adding the absorption due to the various surfaces. For example, suppose we have a room of volume 60 m³, the surfaces of which comprise: 20 m² of plasterboard under joists; 20 m² of thin carpet on wood-board floor; 50 m² of plaster on brick; and 10 m² of window. Also, there are five people seated on wooden chairs. The amounts of absorption (in the 300–600 Hz octave) are then:

	Sabins
Plasterboard: 20 m² @ coeff. of 0·1	2
Carpet: 20 m² @ coeff. of 0·3	6
Plaster on brick: 50 m⁵ @ coeff. of 0·02	1
Windows: 10 m² @ coeff. of 0·1	1
5 people @ 0·4 sabins each	2
Total Absorption	12

In this particular octave band of 300 to 600 Hz the air absorption is negligible, but if the absorption in, say, the 2400–4800 Hz band were required, the above calculations would be repeated with the appropriate coefficients and in addition the amount $x \times V$ would be added. In this band and at a relative humidity of 60% and at a temperature of 20°C, x (from Table XVI) is 0·02, so that the absorption due to the air is 1·2 sabins.

The intensity of the reverberant sound in a room is inversely proportional to the total absorption present, so that the change in the reverberant sound pressure level, in decibels, due to a change in the total absorption present can be got from the intensity relationships shown in Appendix E. Thus doubling the amount of absorption will reduce the reverberant sound level by 3 dB, quadrupling the absorption will reduce the level by 6 dB, and so on.

For example, suppose in the room described above the reverberant sound level in the 300–600 Hz octave is 80 dB. If the 20 m² of plasterboard is replaced by a perforated plasterboard with an absorbent behind it, which has an absorption coefficient of 0·9, the absorption of this new surface will be 18 sabins, i.e. an increase of 16 sabins over the original state. The total absorption has thus gone up from 12 sabins to 28 sabins. The ratio of 28/12 equals 2·3 approximately, which from Appendix E equals an intensity ratio of about 4 dB, so that the reverberant sound level will be reduced to 76 dB.

The amount of absorption in a room can be measured by measuring the reverberation time, as described below.

(d) Reverberation Time

The reverberation time of a room is defined as the time taken for the sound to decrease by 60 dB, after the sound source has stopped. Some estimate of the reverberation time can be made by ear, by making some noise, e.g. a hand-clap, which is about 60 dB louder than the ambient background noise level and measuring with a stop-watch the time taken for the sound to die away to inaudibility. This is obviously a very approximate method and gives no indication of how the reverberation time varies with frequency. A precise measurement is easily made with a level recorder. The sound source can be anything convenient so long as it is loud enough; a pistol is often used, or a loudspeaker emitting a 'white' noise. The measuring microphone, through an octave filter, is connected to a level recorder. If a pistol is used as the source the reading of the level recorder will rise almost instantaneously to a high level and will then drop as the reverberation decreases. Because the reverberation decreases logarithmically, and as the level recorder has a logarithmic response, the decay will be a straight line on the record. For instrumental reasons it is seldom possible to measure the full 60 dB decay, but a 30 dB decay will be sufficient. The actual reverberation time is of course twice the time taken for the 30 dB decay. If a steady sound, e.g. 'white' noise from loudspeakers, is used as a source, then the level recorder trace will be steady while the source is on, and will then show the reverberation time when the source is stopped suddenly. Whether an impulsive or a steady source is used it is better to ignore the first 5 dB or so of the decay. With an impulsive source this is because the level recorder may over-shoot on its way up and will not then be showing the true reverberation time while it recovers from the over-shoot; with a steady sound source the direct sound may be determining the steady reading and when this is removed the level will fall instantaneously before it reaches the true reverberant sound level.

The measurements should be made at several positions in the room and the results averaged.

For dealing with noise problems it is sufficiently accurate to use the following approximate relationship between reverberation time and total absorption:

$$\text{Reverberation time (seconds)} = \frac{0.16V}{A}$$

where V is the volume of the room in cubic metres and A is the total absorption in m² sabins. At the higher frequencies it should be remembered that the term A includes both the surface absorption and the air absorption.

The measurement of the reverberation time is a more accurate way of getting the total absorption in a room than calculating as described above.

For example, if we consider a room of volume 62·5 m³ whose measured reverberation time at 500 Hz is 1 second, the total absorption is $\dfrac{0·16 \times 62·5}{1} = 10$ sabins.

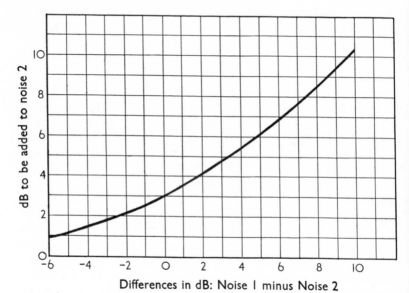

FIG. 89. Addition of Two Pressures

GENERAL

(a) *Addition of Noises*

When it is necessary to consider what the sum of two random noises will be, the intensities and not the pressures must be added. For example, if two equal noises are added the intensity will be double but, as the pressure is proportional to the square root of the intensity, the pressure will go up by a factor of $\sqrt{2}$. Thus if the two pressures are given in decibels the increase will be 3 dB. Fig. 89 can be used to simplify the addition. The difference in dB between the two noises to be added read along the abscissa will give on the ordinate the number of dB to be added to the smaller noise to give the total noise. For example, if two noises have pressure levels of 80 and 85 dB respectively, the difference is 5 dB which, from Fig. 89, indicates that 6 dB should be added to the smaller noise, so the combined noise level is 86 dB.

Similar considerations apply when, as happens in some instances, it may be desirable to calculate from an octave analysis the total sound pressure over the whole frequency range. To do this the intensities in each octave band must be added. For example, the octave-band levels of Fig. 90 are given

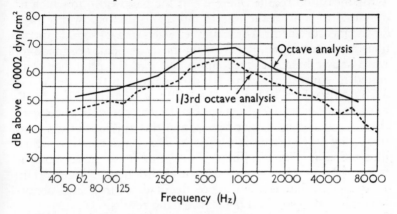

FIG. 90. Analysis of Noise in a Canteen

in Table XVII, and the intensities in each octave band relative to some convenient reference (in this case 60 dB) are calculated from intensity ratios on the dB table on p. 321. The sum of all these relative intensities is 14·15, and this is 11·5 dB above the reference of 60 dB. The total level over the whole frequency range is therefore $60 + 11·5 = 71·5$ dB.

TABLE XVII

Octave Band (Hz)	Sound Pressure Levels (dB above 0·0002 dyn/cm²)	Intensity Relative to a Level of 60 dB
37–75	52	0·15
75–150	54	0·25
150–300	59	0·80
300–600	67	5·0
600–1200	68	6·3
1200–2400	61	1·25
2400–4800	55	0·30
4800–9600	49	0·10
		Sum 14·15

(b) Effect of Bandwidth

For all noises except those consisting of a few discrete frequencies there will obviously be more energy in a wide pass-band than there will be in a narrow pass-band. Thus in any presentation of results it is essential to say what pass-band the sound pressures were measured in. Occasionally, it is desirable to transfer results obtained in one set of pass-bands to another set. The conversion is made by adding $10 \log \Delta f_1/\Delta f_2$ (in dB) to the measured results where Δf_1 is the bandwidth to which it is desired to convert the results and Δf_2 is the bandwidth used for the measurements. For example, when impact sound insulation is measured (see p. 275) if the measurements are made in 1/3rd octave bands it is recommended that the results be converted to equivalent octave band pressures. In this case $\Delta f_1 = 3 \times \Delta f_2$, therefore $10 \log \Delta f_1/\Delta f_2 = 10 \log 3 = 5$ dB. Thus 5 dB must be added to 1/3rd octave band measurements to convert them to equivalent octave-bands, and this applies to the whole frequency range because the ratio $\Delta f_1/\Delta f_2$ is the same. A further illustration is Fig. 90 which shows the noise in a canteen measured in octave bands and in 1/3rd octave bands. The 1/3rd octave bands show more detail of course, and are in general 5 dB lower than the octave band results.

(c) Calculation of Loudness

The loudness level of a noise in phons can only be obtained, strictly speaking, by comparing subjectively the noise with the standard tone of 1000 Hz (see Chapter 1). This technique requires a number of observers operating under controlled conditions, and is not practicable in the field. One early attempt to measure loudness objectively (i.e. to be able to read the loudness off a meter) is the 'sound-level meter'. This consists of a measuring microphone, amplifier and meter and it incorporates three 'weighting networks' (Fig. 91) which approximate the frequency responses of the ear. The original idea was that the various networks should be switched in depending on the loudness of the noise. This did not work, but it has been found recently that if the 'A' weighting response is used, *for all loudness levels*, then for several types of noise the reading of the meter can be converted into loudness (in 'Stevens' phons, defined below) with accuracy sufficient for most practical purposes by multiplying the reading by 1·05 and adding on 10. Thus, for example, an 'A' weighting reading of 80 dB indicates a loudness of 94 phons. A modern precision sound level meter is shown in Plate XIV.

The types of noises for which this relationship holds are in general noises which are spread over a fairly wide range of frequencies, e.g. traffic noises. It may be considerably wrong for noises which contain strong discrete frequencies, e.g. a band-saw.

A more reliable method of measuring the loudness of noise in the field, and one which is adequate for all practical purposes,

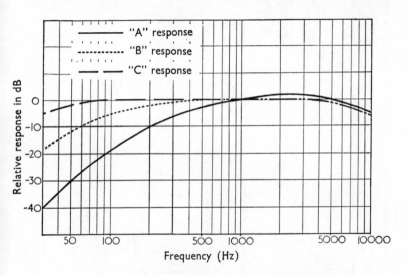

Fig. 91. 'A', 'B' and 'C' ('40 dB', '70 dB' and 'Flat' Weighting) Responses of Sound-level Meter

is to analyse the noise into octave bands, to calculate from the sound-pressure level in each band the loudness of each band, and then to sum these loudnesses to get the total loudness. Various methods of summation are in use; the one given here is due to S. S. Stevens.

The exact procedure is as follows. From Fig. 92 the loudness of the noise in each octave band (in sones, see p. 35) is obtained. The total loudness in sones is calculated from

$$S_T = S_M + 0·3(\sum S - S_M) \quad . \quad . \quad . \quad (10)$$

263

where S_T is the total loudness in sones, S_M is the loudness in sones of the loudest octave band (whichever it turns out to be for a particular noise) and $\sum S$ is the sum of the loudnesses of all

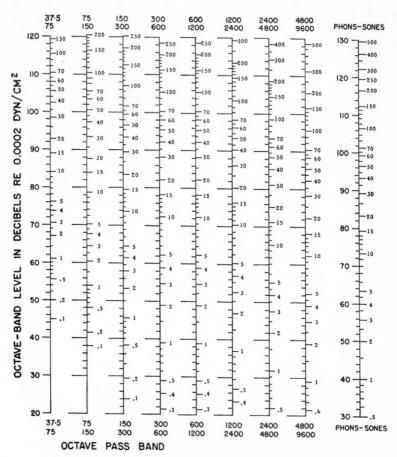

FIG. 92. Sound-pressure Levels and Sones in Octave Bands

the bands. Finally, S_T in sones is converted to phons using the right-hand side column of Fig. 92.

The graph (Fig. 90) showing the noise in a canteen can be used as an example. Table XVIII shows the process of calculation.

TABLE XVIII

Octave Band (Hz)	Sound Pressure Level (dB above 0·0002 dyn/cm²)	Loudness in Sones of Each Band (from Fig. 92)
37–75	52	0·3
75–100	54	1·0
150–300	59	3·4
300–600	67	7·0
600–1200	68	7·5
1200–2400	61	5·5
2400–4800	55	5·0
4800–9600	49	4·0
		33·7

$S_M = 7·5$, therefore relationship (10) is:

$$S_T = 7·5 + 0·3 \ (33·7 - 7·5)$$
$$= 15·4 \text{ sones}$$
$$= \textbf{80 phons}$$

INSULATION IN BUILDINGS

By air-borne sound insulation (as distinct from impact sound insulation, discussed below) is meant the insulation of sounds originating in the air, e.g. voices. It is also sometimes called structure-borne sound insulation, because the sound travels via the structure and not via air-paths.

The sound transmission coefficient, $\mathscr{T}$, of a partition (we will use the term partition generally; i.e. to include walls, floors, roofs, windows, etc.) is the ratio of the sound energy transmitted through it to the sound energy incident on it. In decibels, the sound reduction index, R, is $10 \log (1/\mathscr{T})$. Thus, for example, if a partition radiates on one side one five-hundredth of the energy incident on the other side, $\mathscr{T}$ is 0·002 and R is 27 dB.

The air-borne sound insulation of any partition varies with frequency and it is therefore necessary to measure it over a frequency range. This range should cover all frequencies likely to be important, but there are practical limitations. The lower frequency limit is set by the fact that the wavelength of the sound may be the same order of size as the dimensions of the room being measured. For example, at 100 Hz the wavelength is 3·35 m, and this means that when noise at this frequency is generated in a room there will be big differences in pressure levels from one part of a room to another, due to the standing wave patterns that are set up. It is then difficult to decide what the average pressure levels in the rooms are. Further, the partition may itself be about the same size as the wavelength (except in thickness) and this makes its sound insulation behaviour erratic. On the other hand, these low frequencies are often important subjectively, that is to say it is often low-frequency noises which are the loudest and most annoying of the noises which come through a wall or floor. A reasonable compromise is 100 Hz and this is the lowest frequency at which sound insulation measurements are usually made.

At the other end of the frequency scale, there is no point in going up to the highest audio-frequencies because it is very seldom that these frequencies are subjectively important. Again, a reasonable compromise for the upper limit is 3150 Hz (i.e. 5 octaves above 100 Hz). This upper limit may be a little high for partitions with good insulation because the insulation is usually so great as to reduce many ordinary noises, e.g. voices, to inaudibility above about 2000 Hz. On the other hand, partitions with low insulation may transmit appreciable amounts of noise at higher frequencies than this. So as to avoid having different ranges for partitions of different insulations, this one range of 100 to 3150 Hz is used for all measurements.

Laboratory measurements of sound insulation, and field measurements between two reasonably reverberant rooms (i.e. all rooms in dwellings, offices, schools, hospitals and most rooms in factories) are made in the following manner. (See British Standard 2750:1956 for a complete statement.) Warble tones are fed to loudspeakers in one room (the source room). The aim is to make the sound field in this room as diffuse as possible, and usually two loud-speakers are used. The sound pressures are measured at several positions in the room. This is necessary because the sound field is never completely diffuse, and the average pressure throughout the room must be obtained. The readings from the several positions are averaged to get the average pressure level (L_1) in the room. The average pressure level should, strictly, be obtained from arithmetic mean of the intensities at each position at each frequency. This arithmetic takes some time and it is usually sufficiently accurate to take the arithmetic mean of the decibel values. The errors introduced will apply to both source and receiving rooms and will therefore tend to cancel out.

The measurements are made at 1/3rd octave intervals from 100 Hz up to 3150 Hz, i.e. at 100, 125, 160, 200, 250, 315, 400, 500, 630, 800, 1000, 1250, 1600, 2000, 2500 and 3150 Hz. (As an alternative to warble tones, white noise may be used for the source with the appropriate filters introduced into the circuits.)

After the pressures have been measured in the source room the microphones are moved to the room on the other side of the partition being measured (the receiving room). The same voltages at the same frequencies are now fed to the loudspeakers so as to reproduce the same sound pressures as before in the source room. The sound pressures in the receiving room are measured at several positions, and the average pressure obtained (L_2).

The level, L_2, in the receiving room will depend not only on the sound reduction index of the partition, but also on the area of the partition and on the amount of absorption in the receiving room. In the laboratory, precautions are taken so that flanking transmission is negligible, and the sound reduction index is then given by:

$$R = L_1 - L_2 + 10 \log S - 10 \log A \quad . \quad . \quad (11)$$

where S is the area in m² of the partition being measured and A is the absorption in m² units (sabins) in the receiving room. The quantity A is obtained by measuring the reverberation time of the receiving room and using the usual Sabine formula. Thus:

$$A = \frac{0 \cdot 16 \times \text{volume of receiving room (in m}^3)}{\text{Reverberation time}} \quad (12)$$

There are two minor differences between laboratory and field measurements of R. The first is that if the size of the partition being measured in the field is very different from the laboratory specimen then its sound-insulating behaviour will be different. If, however, the laboratory specimen is a 'reasonable' size, let us say with a minimum dimension of 2 m, then this difference will not usually be great. Secondly, the edge-fixing conditions between field and laboratory will probably be different, and this may have some effect, depending on the construction.

But the major difference between laboratory and field measurements is that in the laboratory indirect transmission is always negligible but in the field it may predominate. An outstanding example of this is the joist floor with a floating raft and with pugging. Measured in the laboratory this floor has an insulation of the order of 65 dB (averaged over the frequency range 100 to 3150 Hz). But if this floor is used in practice on 110 mm supporting walls (i.e. the inner leaf of an 280 mm cavity wall) then the insulation is only about 45 dB.

There are three classes of sound-insulation problems in the field. The first is the measurement of the existing sound insulation; the second is a similar measurement but with a subsequent correction for a change of conditions in the receiving room; the third is a calculation of the insulation, knowing the particular arrangement of the rooms and the sound-reduction factor of the partition or partitions concerned.

In neither of the first two classes is there any difficulty with indirect transmission because it is obviously part of the measured value.

In the first type of problem—a measurement of the existing conditions—the insulation is simply given by $L_1 - L_2$, where L_1 and L_2 are, as before, the measured average levels in the source room and the receiving rooms respectively. For the second class of problem—where it is necessary to correct the measured value (i.e. $L_1 - L_2$) to some other value to allow for a change in the conditions in the receiving room—two methods of correction may be used, as appropriate.

The first alternative method of correction is to use a reference reverberation time for the receiving room. The corrected value of insulation in decibels is then given by:

$$D = L_1 - L_2 + 10 \log T - 10 \log T_{\text{REF}} \quad . \quad (13)$$

where L_1 and L_2 are as before, T is the measured reverberation time of the receiving room and T_{REF} is the reference reverberation time.

The second alternative method is to use a reference absorption of $10 \, \text{m}^2$ sabins in the receiving room. The corrected value of insulation is then given by

$$D = L_1 - L_2 + 10 \log 10 - 10 \log A \quad . \quad (14)$$

where A is the measured absorption in the receiving room.

In dwellings it has been found that the reverberation time of furnished living-rooms is usually about 0·5 second, independent of their volume. This means that, if the measured values are corrected to a standard reverberation time of 0·5 second, the corrected measured values will be close to the values existing in practice when the rooms are furnished and occupied.

The normalised level difference in decibels is then given by:

$$D_N = L_1 - L_2 + 10 \log T - 10 \log 0.5 \qquad . \qquad (15)$$

where L_1 and L_2 are the average sound-pressure levels in the source and receiving rooms respectively, and T is the reverberation time of the receiving room under the conditions of measurement.

The laboratory relationship (11) can be useful for comparing field results with laboratory results on a given wall or floor.

For example, suppose the sound insulation of a floating concrete floor is to be measured. It would be done at all frequencies from 100 Hz to 3150 Hz but we will consider one frequency only, 250 Hz. The rooms are unfurnished. The decibel readings at six positions in the source room may be: 99, 100, 95, 95, 97 and 98, the average level L_1 (if we average the decibel readings and not the powers) being 97. In the receiving room the readings are: 60, 55, 59, 59, 63 and 59; the average, L_2, is 59. Thus the difference under the conditions of measurement, i.e. unfurnished, is 38 dB. The measured reverberation time in the receiving room is 1·53 seconds, therefore D_N, from relationship (15) is:

$$D_N = 97 - 59 + 10 \log 1.53 - 10 \log 0.5$$
$$= 38 + 5$$
$$= \mathbf{43 \ dB}$$

Note that if the 'true' L's had been calculated from the average of the powers then L_1 would have been 98 dB and L_2 would have been 60 dB, thus leaving D_N unchanged.

To compare this field result with a laboratory measurement of the same floor, we need to know the area of the floor (S) and the volume of the receiving room. Let these be 12 m^2 and 31·5 m^3 respectively. The total absorption, A, in the receiving room is obtained from relationship (12):

$$A = \frac{0.16 \times 31.5}{1.53}$$
$$= 3.3 \text{ sabins}$$

Therefore R, from relationship (11) is:

$$R = 97 - 59 + 10 \log (12/3.3)$$
$$= 38 + 5.5$$
$$= \mathbf{43.5 \ dB}$$

Thus in this example the laboratory and field corrections are practically the same.

The third class of problem—calculations of the expected sound insulation—may be affected by indirect transmission. If it can be assumed that the indirect transmission is negligible, then the sound-reduction factor as measured in the laboratory can be used in relationship (16) to give the expected difference in levels between the two rooms. Thus:

$$(L_1 - L_2) = R - 10 \log (S/A) \qquad . \qquad . \qquad (16)$$

where S is the area of the dividing partition and A is the absorption in the receiving room.

For example, suppose that a partition is to be used between two offices and that R for this partition at a certain frequency is known to be 30 dB. Then, if the area of the partition is 8 m^2 and the absorption in the receiving room is 20 sabins:

$$(L_1 - L_2) = 30 - 10 \log (8/20)$$
$$= 30 + 4$$
$$= \textbf{34 dB}$$

If the absorption in the receiving room is not known an approximate amount can be obtained in the case of lightly furnished rooms (i.e. offices, hospital wards, school classrooms) by assuming a reverberation time of one second. A will then be given by $0 \cdot 16 \times$ volume of receiving room.

If the indirect transmission is not negligible then accurate calculations cannot be made. Instead, if some sort of calculation must be made, we can only use field measurements of the same sort of construction under the same sort of flanking conditions to get some estimate of R including the indirect transmission. Appendix C gives the values of R for several types of construction, and included in it are some estimates for R including the indirect transmission. These estimated values only apply under the conditions stated.

If the partition between two rooms is made up of more than one type, then the composite sound-reduction index is obtained from the average sound-transmission coefficient, calculated as follows (assuming each element of the construction is subject to the same sound-pressure level):

$$\mathscr{T}_{\text{av}} = \frac{\mathscr{T}_1 S_1 + \mathscr{T}_2 S_2 + \mathscr{T}_3 S_3 + \ldots}{S} \qquad . \qquad (17)$$

where $\mathscr{T}_1, \mathscr{T}_2, \mathscr{T}_3 \ldots$ are the coefficients for each element; $S_1, S_2, S_3 \ldots$ are their corresponding areas; and S is the sum of all these areas. The average sound-reduction index is then $10 \log (1/\mathcal{T}_{av})$.

For example, suppose a partition between two rooms is made up of a 7·5 m² area of a material whose sound-reduction index, at a certain frequency, is 30 dB, and a 2·5 m² area of another material whose sound-reduction index is 10 dB. Then:

$$30 = 10 \log \frac{1}{\mathcal{T}_1} \quad \text{and} \quad 10 = 10 \log \frac{1}{\mathcal{T}_2}$$

Thus $\mathcal{T}_1 = 0·001$ and $\mathcal{T}_2 = 0·1$; $\mathcal{T}_1 S_1 = 0·0075$ and $\mathcal{T}_2 S_2 = 0·25$.

$$\mathcal{T}_{av} = \frac{0·0075 + 0·25}{10} = \frac{0·2575}{10} = 0·02575$$

$$\therefore R_{av} = 10 \log \frac{1}{0·02575} = \mathbf{16\ dB}$$

A graph for the rapid calculation of the composite insulation of a partition made up of two areas is shown in Fig. 93. The ordinate shows the ratios of the two areas and the abscissa the loss of insulation, i.e. to be subtracted from the higher value. In the previous example the ratio of areas was 1/3rd; the difference between the two insulations was 20 dB; so from Fig. 93 the loss of insulation is seen to be 14 dB. The actual insulation is thus $30 - 14 = 16$ dB.

We have so far been considering only the insulation between two 'reasonably reverberant' rooms. ('Reasonably reverberant' is difficult to define, but as a rough working rule we can say that a room will be sufficiently reverberant if the total absorption in the room (in sabins) is not less than $4l^2$ where l is the linear dimension of the room normal to the partition, i.e. horizontally when a wall is involved and vertically when a floor is concerned.) If the rooms are not reasonably reverberant then measurements and calculations become rather involved, and as such rooms are rare we will not deal with them here.

INSULATION BETWEEN INSIDE AND OUTSIDE OF BUILDINGS

The other practical cases are the insulation from outside to inside a building, and vice-versa. In such cases it is nearly always a question of calculating L_2 (the level at the receiving position) from a knowledge of L_1 (the level due to the source) and of R, rather than of obtaining R from measurements of L_1 and L_2; thus the relationships given below show how to calculate L_2, and cannot always be used to measure R.

The first case is the insulation from a reverberant room to outside. If L_1 is the average level (in decibels) in the reverberant room and L_2 is defined as the level very close to the outside of the partition then:

$$L_2 = L_1 - R - 6 \qquad . \quad . \quad . \quad (18)$$

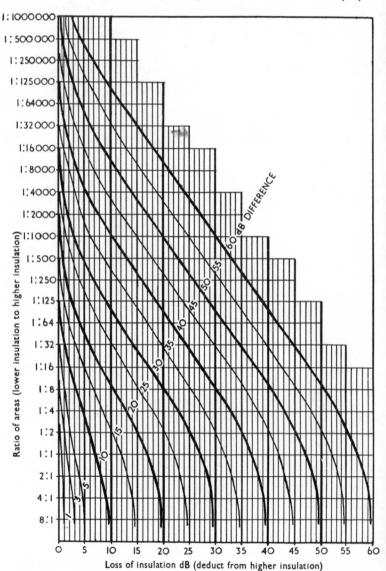

FIG. 93. Composite Insulation of Two Areas

For indoors-to-outdoors calculations we usually want to know the level at a certain distance from the partition. An exact relationship cannot be given, but approximately:

$$L_2 = L_1 - R + 10 \log S - 20 \log r - 14 \quad . \quad (19)$$

where L_1 is the average level in the reverberant room, L_2 is the level at a distance r (in m) from the partition along a line normal to the partition (r is assumed large compared with the size of the partition), and S is the area of the partition (in m^2).

The simplest example is when a building has one obvious 'weak link', e.g. a window in a brick wall, and it is required to know the noise level at some distance from this window and opposite to it. Relationship (19) can then be directly applied. Conditions are a little more complicated if there is more than one path for the sound to travel from inside to outside, e.g. a window and a door, but it is only necessary to calculate two separate values of L_2 from the two values of R and S and then to add these two values (as described on p. 260).

Conditions become much more complicated when the noise level is required at a position not on the normal to the radiating area, and in fact the level cannot be calculated with any certainty. As a rough guide, Table XIX shows the reduction in level expected at positions at 90° to the normal compared with the level on the normal (based on empirical data for horizontal openings given by Dyer), and applies to openings. These figures apply within 1 or 2 dB for distances of up to 300 m. At greater distances the reductions will be the same up to about the 300–600 Hz octave, but at higher frequencies will not exceed 8 to 10 dB.

Similar information about the radiation by walls or roofs is not available.

TABLE XIX

REDUCTION (dB) AT SIDE POSITION COMPARED WITH NORMAL POSITION

Octave Band (Hz)	Area of Opening	
	10 m^2	50 m^2
37–75	0	0
75–150	0	2
150–300	1	3
300–600	4	7
600–1200	7	10
1200–2400	11	13
2400–4800	13	15
4800–9600	15	17

Secondly, for calculations concerning outside-to-inside conditions the relationship is:

$$L_2 = L_1 - R + 10 \log S - 10 \log A + 6 \quad . \quad (20)$$

where L_1 is the sound level expected to be present at the position of the partition and coming from a direction normal to the partition, L_2 is the reverberant sound level inside the room, S is the area of the partition and A is the absorption in the room. As for the inside-to-outside case, if the surface of the room facing the noise is made up of two or more parts (e.g. a window and a door in a brick wall) then separate values of L_2 are calculated and the intensities added as described on p. 260.

We can make some approximate estimates of the sound pressures on the surfaces of a building not facing the noise source, assuming that the distance to the noise source is large compared with the building dimensions. Consider a rectangular building with the noise coming horizontally towards it. Then for the side facing the noise, relationship (20) as it stands should be used. For the sides parallel to the direction of the noise 6 dB should be subtracted from L_1 (assuming the dimensions of the building are large compared to the wavelength, which they usually will be). For the roof, if it is flat then again 6 dB should be subtracted from L_1; if it is pitched 3 dB should be subtracted for the side facing towards the noise. For the side of the building facing away from the noise, the amount to be subtracted from L_1 will depend on the size of the building and can not be assessed accurately. However, it will probably be at least 10 dB at the lower frequencies.

If the sound is coming vertically, i.e. from an aircraft in flight, then for a flat roof relationship (20) is used; for a pitched roof 3 dB is subtracted from L_1, and for all the walls 6 dB is subtracted from L_1.

ENCLOSURES

It is sometimes convenient to calculate what the reduction in sound level will be when a source of noise is enclosed. First, we will consider a noise source out of doors. If the source is directional then an accurate calculation is complicated but, assuming that 60% of the power is radiated in one direction, then the reduction in level due to completely enclosing the source is given by:

$$L_1 - L_2 = R - 10 \log S + 10 \log A + 2 \quad . \quad (21)$$

where L_1 is the level at a distance from the source on the axis of main sound radiation and before enclosure, L_2 is the level at the same position after enclosure, R and S are the sound-reduction index and the area respectively of that part of the enclosure facing towards the measuring position and A is the absorption in the enclosure. It is assumed that the insulation of the rest of the enclosure is of the same order as the part facing the measuring position.

Note that if A is small the change in level due to the enclosure will be small.

At positions not on the axis of main radiation the reduction in level will not be as much as indicated by relationship (21).

Secondly, if the noise source is in a room, the difference in the reverberant sound level in the room before and after the noise source is enclosed is given by:

$$L_1 - L_2 = R - 10 \log S + 10 \log A \quad . \quad . \quad (22)$$

where R is the average sound-reduction index of the enclosure, S is the corresponding area and A is the absorption inside the enclosure.

Again, note that if A is small the change in level will be small.

IMPACT SOUND INSULATION

By impact sound insulation is meant the insulation against noises generated by impacts on the structure, e.g. footsteps. (Vibration as defined on p. 187 is not included here.) Most impacts occur on a floor, and we will not deal with the comparatively rare problem of impacts on a wall.

Impact sound insulation is measured using a standard 'impact' or 'footsteps' machine. This produces impacts of standard energy on the surface of the floor being measured, and the resulting sound pressures in the room below (the receiving room) are measured. It should be noted that unlike air-borne sound insulation measurements which are only relative, an absolute measure of sound pressure is now required, needing a calibrated microphone. The standard impact machine consists of five hammers with metal heads each weighing 500 grammes falling freely through 4 cm at a total rate of striking of 10 blows per second. This makes more noise than most impact noises met with in practice, but this high noise level is necessary for reliable measurements (e.g. to ensure that the noise being measured is well above any other noise that may be present). The fact that it is higher does not matter because all that is required is a comparison between floors. One design of the standard machine is shown in Plate XIII.

The pressure-level measurements in the receiving room are made at several positions and the results averaged. As with airborne sound insulation, the average of the intensities should be taken but if, to save time, the decibel readings are averaged, 1 dB should be added to the average if the spread of the readings is of the order of 10 dB.

The sound pressure in the receiving room is analysed into frequency bands, and, while 1/3rd octave analysis is desirable (over the range 100 to 3150 Hz), octave band analysis will do if 1/3rd octave filters are not available. For ease of comparison, the results are best plotted at 1/3rd octave intervals, even if octave filters are used. The measured level of a noise will depend on the width of the filter used for analysis, and in impact sound insulation it is customary to refer all measurements to octave band widths. Thus if 1/3rd octave filters are used, 5 dB is added to the levels obtained.

The differences between laboratory and field measurements tend to be greater for impact sound insulation than they are for air-borne sound insulation. This is because not only is the indirect transmission different but the size of the floor and the edge conditions have a greater effect. Further, the laboratory method of correction (see below) makes no allowance for size. For these reasons, it is often difficult to estimate what will happen in the field from laboratory measurements. What can most usefully be done in the laboratory is to compare the effect of modifying a floor, e.g. by putting a soft layer on top of it.

Measured values of impact sound insulation are corrected in the laboratory or in the field by reference to a standard amount of absorption (10 m^2 sabins) in the receiving room. Thus, the normalised impact sound transmission level (in decibels) is given by:

$$L_N = L + 10 \log A - 10 \log 10 \qquad . \quad (23)$$

where L is the measured level and A is the measured absorption in the receiving room.

An alternative method in the field is to correct to a standard reverberation time. Thus:

$$L_N = L + 10 \log T_{REF} - 10 \log T \qquad . \quad (24)$$

where L is the measured level, T is the measured reverberation time in the receiving room and T_{REF} is the reference reverberation time.

For dwellings, T_{REF} is taken as 0·5 seconds, and the grading system described on p. 297 applies to measurements corrected in this way.

For schools, offices and hospital wards T_{REF} will be taken as 1 second (in the absence of any firmer information) and the recommendations made for these buildings in Chapter 8 apply to measurements or estimates corrected in this way.

VENTILATION PLANTS

The noise problems associated with ventilation plants are described in Chapter 8. The calculations that can be made are given here, and comprise (*a*) attenuation of sound along ducts, (*b*) noise made by fans, (*c*) noise made by grilles, and (*d*) applications. More rigorous treatment than that given here is possible and this will be found in papers published by Beranek *et al.* and in the *Handbook of Acoustic Noise Control*, Supplement 1.

(*a*) *Attenuation of Sound along Ducts*

If the internal surfaces of a duct are not sound absorbent then sound will travel along them for long distances with little reduction in intensity. If the internal sides of the ducts are lined with an absorbent material then the attenuation is given approximately by:

$$R = \frac{\alpha^{1·4} . P}{S} \qquad . \quad . \quad . \quad . \quad (25)$$

where R is the attentuation in dB per metre, P is the perimeter of the duct in metres, S is the cross-sectional area of the duct in square metres and α is the absorption coefficient of the lining. ($\alpha^{1·4}$) is plotted against α on Fig. 94.

This relationship is accurate enough for most practical purposes when the area of the duct does not exceed about 0·3 m², when the ratio of the dimensions of the duct sides does not exceed 2:1, and for the usual kinds of absorbent material.

The attenuation varies with frequency because the sound absorption coefficient is frequency dependent. The above relationship tends to underestimate the attenuation at low frequencies and to overestimate it at high frequencies. However, this does not necessarily mean that there will be too much attenuation at the low frequencies, because the absorption coefficient of the lining material will usually be small at the low frequencies.

For larger duct areas than 0·3 m², or when the ratio of the sides exceeds 2:1, it is desirable to use splitters consisting of an impervious core lined on both sides with absorbent, the effect of which is to increase P without appreciably affecting S.

Little is known about the attenuation provided by bends in lined ducts, but 90° bends will provide the approximate attenuations shown in Table XX.

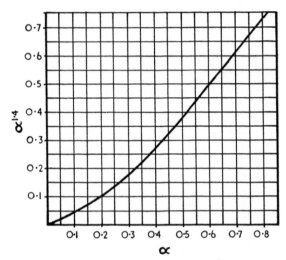

Fig. 94. $\alpha^{1·4}$ as a Function of α

TABLE XX

ATTENUATION PER 90° BEND (dB) IN SQUARE, LINED DUCTS

Octave Band (Hz)	Duct Width	
	0·3 m	1·2 m
37–75	0	1
75–150	1	3
150–300	3	8
300–600	6	16
600–1200	8	17
1200–2400	16	18
2400–4800	17	18
4800–9600	18	18

When the cross-sectional area of a duct changes the change in the sound level is given by:

$$\text{Change in dB} = 10 \log \frac{S_1}{S_2} \qquad . \quad . \quad (26)$$

where S_1 is the area before the change and S_2 is the area afterwards. Thus, for example, if the area of a duct changes from 1 m² to 0·5 m², then the level increases by $10 \log (1/0 \cdot 5) = +3$ dB.

If the duct opens into a large volume containing absorbent and then continues on, the attenuation introduced (provided the exit duct is not opposite the inlet duct) is given approximately by:

$$\text{Attenuation} = 10 \log \frac{A}{S} . \quad . \quad (27)$$

where A is the absorption in the volume and S is the duct area.

Immediately outside the outlet of a duct the sound-pressure level will be the inlet sound-pressure level minus the attenuation of the duct. If the duct opens into a room the sound level at positions not close to the duct outlet will be less than this, depending on the area of the duct opening and the amount of absorption in the room. Usually it is this reverberant sound level which is of most interest, but it should be remembered that the level just outside the duct exit will be considerably (10 to 20 dB) higher.

The relationships (some of them approximate) for various conditions are given below.

(i) Duct between two reverberant rooms:

$$L_2 = L_1 - R + 10 \log S - 10 \log A \qquad . \quad (28)$$

where L_1 and L_2 are the reverberant sound levels in the source and the receiving rooms respectively, R is the total attenuation of the duct, S is the area of the duct opening (in square metres) and A is the absorption in the receiving room (in m² sabins).

(ii) Duct between a reverberant room which is the source of the noise and the open air, with the open air end several metres away from all reflecting surfaces:

$$L_2 = L_1 - R + 10 \log S - 20 \log r - 17 \qquad . \quad (29)$$

where L_1 is the reverberant sound level in the room, L_2 is the level at a distance r (in metres) from the duct exit and on the axis of the duct, and R and S are as before.

(iii) Duct between a reverberant room which is the source

of the noise and the open air, with the open air end flush with the exterior surface:

$$L_2 = L_1 - R + 10 \log S - 20 \log r - 14 \quad . \quad (30)$$

where all symbols are as before.

(These last two relationships apply to the lower frequencies. At the higher frequencies L_2 will be rather higher due to the 'beaming' of the sound, but in most practical cases it is the lower frequencies which are most important.)

(iv) Duct between the open air where the noise source is, and a reverberant room:

$$L_2 = L_1 - R + 10 \log S - 10 \log A + 6 \quad . \quad (31)$$

where L_1 is the level in the open air near to the duct entrance, L_2 is the reverberant sound level in the room and the other symbols are as before.

Note that if more than one duct is contributing to the noise the intensities of the noise must be added as described on p. 260.

Noise Generated by Ventilation Fans

We are dealing here only with the aerodynamic noise generated by fans, it being assumed that the mechanical noises are negligible.

The total (i.e. full frequency range) sound-pressure level at either the inlet or outlet of a fan (assuming that the areas of the two ducts are equal) and when the fan is running at near its maximum rated horsepower is given approximately by:

$$L_T = 97 + 10 \log \text{h.p.} - 10 \log 10 \cdot 8\, S \quad . \quad (32)$$

where h.p. is the rated horsepower of the fan and S is the area of either duct in square metres.

If the air-pressure drop (P, in mm of water) across the fan exceeds 25, then $10 \log P/25$ should be added to the right-hand side of this relationship.

Fig. 95 shows the sound-pressure levels in the various octave bands for two types of fan relative to the total level, so when L_T is known from relationship (32) the level in any octave band existing at the entrance to the duct can be obtained from Fig. 95.

For a particular fan run at various speeds the pressure level generated varies as $20 \log \text{h.p.}$ or $50 \log \text{r.p.m.}$ Thus if a fan is run at one-half its rated value the level will be reduced by 6 dB.

This compares with 3 dB per doubling of the rated horsepower shown by relationship (32). In other words, a large fan run at half its rating will produce 3 dB less noise than a fan of half the rating run at full power.

When the noise level L_1 in a particular octave band at the entrance to the duct has been obtained the following relationships apply:

(i) When duct goes to a reverberant room:

$$L_2 = L_1 - R + 10 \log S - 10 \log A + 6 \quad . \quad (33)$$

where L_2 is the level of the reverberant sound in the room, R is the total attenuation of the duct, S is the area of the duct opening and A is the absorption in the room.

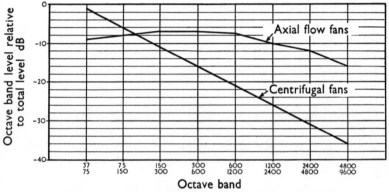

FIG. 95. Octave Levels Relative to Total Levels for Fan Noise

(ii) When duct goes to the open air and when the open air end is away from all reflecting surfaces:

$$L_2 = L_1 - R - 20 \log r + 10 \log S - 11 \quad . \quad (34)$$

where L_2 is the level at a distance r.

(iii) When duct goes to the open air and when the open air end is flush with the surface of the building:

$$L_2 = L_1 - R - 20 \log r + 10 \log S - 8 \quad . \quad (35)$$

(These two relationships apply to the lower frequencies. At the higher frequencies L_2 will be rather higher due to the 'beaming' of the sound, but in most practical cases it is the lower frequencies which are most important.)

281

GRILLE NOISE

Some noise will be generated by the air passing through any grille across the exit of the duct. This is seldom important, but a very rough indication of the level in the room is given by:

$$L_T = 80 + 10 \log S - 10 \log A + 10 \log P/25 \quad . (36)$$

where L_T is the total reverberant sound level, S is the area of the duct, A is the absorption in the room and P is the air-pressure drop across the grille in millimetres of water.

From L_T the levels in the various octave bands are obtained from Fig. 96.

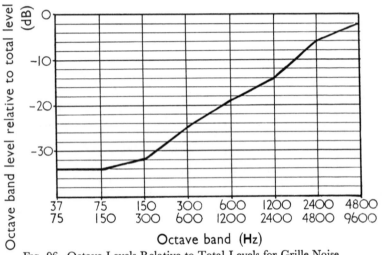

FIG. 96. Octave Levels Relative to Total Levels for Grille Noise

If there is more than one grille the several values of L_T must be added as described on p. 260, e.g. a 3-dB increase for two grilles of equal loudness.

EXAMPLE

Suppose a ventilation plant is arranged as shown in Fig. 97 and it is desired to calculate the reverberant sound level in the room due (a) to the external noise entering via the plant, and (b) to the plant itself. The calculations will be done for the 300–600 Hz octave only.

Let the relevant conditions be as follows:

Sound level just outside the duct entrance: 80 dB (in the 300–600 Hz band).

282

Duct between outside and fan untreated.

Centrifugal fan 20 h.p.

Outlet duct from fan 2·4 m × 0·6 m with splitters so that it is divided into four areas 0·6 × 0·6 m.

Duct and splitters lined with material of absorption coefficient 0·65 (in the 300–600 Hz band).

Length of duct between fan and room 6 m, with one 90° bend.

Pressure drop across exit grille, 0·25 mm of water.

Absorption in room 20 sabins.

Then:

Attenuation along duct (relationship (25), p. 277)

$$= \frac{0{\cdot}65^{1{\cdot}4} \times 2{\cdot}4}{0{\cdot}36} \text{ dB/m}$$

$$= 3{\cdot}6 \text{ dB/m}$$

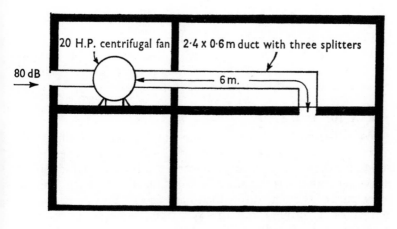

FIG. 97. Example of Ventilation Plant

Therefore total attenuation, R, along 6 m run of duct $\simeq$ 22 dB.

The attenuation due to the 90° bend at this frequency is, from Table XX, 10 dB (interpolation).

Thus, from relationship (31), the reverberant sound level in the room due to the external 80 dB

$$= 80 - 22 - 10 + 10 \log 1{\cdot}44 - 10 \log 20 + 6 \text{ dB}$$

$$= \textbf{42 dB}$$

From relationship (32), the total sound level due to the fan at the entrance to the duct is:

$$L_T = 97 + 10 \log 20 - 10 \log 16 \text{ dB}$$
$$= \textbf{98 dB}$$

From Fig. 95 the level in the 300–600 Hz band is 16 dB less than this, i.e. 82 dB.

Thus from relationship (33) the reverberant sound level in the room is:

$$L_2 = 82 - 32 + 10 \log 1{\cdot}44 - 10 \log 20 + 6 \text{ dB}$$
$$= \textbf{44 dB}$$

The total reverberant grille noise (from relationship 36) is:

$$L_T = 80 + 10 \log S - 10 \log A + 10 \log P/25 \text{ dB}$$

and, from Fig. 96, the L in the 300–600 Hz band is 25 dB less than this. So the grille noise (from relationship) is:

$$80 + 10 \log 1{\cdot}44 - 10 \log 20 + 10 \log 0{\cdot}01 - 25 = \textbf{23 dB}$$

Thus the total reverberant level in the room in the 300–600 Hz band is the sum of the three intensities corresponding to 42, 44 and 23 dB. The 23 dB part is negligible, and the sum of 42 and 44 dB is, approximately, 46 dB.

Vibration Isolation

We will deal here only with the simplest form of vibration, namely vibration with one degree of freedom and in the steady state. Further, we are only concerned with the reduction of noise generated not by the vibrating machinery itself but by some part of the structure which has been set into vibration by transmission through the structure. For example, a rotating machine in the basement of a building will transmit vibrations to the structure which will travel to other parts of the building and there set walls, floors or other areas into vibration, and thus radiate noise.

There is no way of calculating the amount of noise that will be generated in such a way. What can be done is to calculate the reduction, R, of noise level in a given case that will be effected by the use of a resilient mounting to the offending machine. This is given approximately by:

$$R = 40 \log \left(\frac{f}{f_0}\right) \text{ dB} \quad . \quad . \quad . \quad (37)$$

where f is the frequency of the driving force and f_0 is the natural frequency of the mounting, and where $f/f_0 \gg 1$. If f/f_0 is not much greater than unity the above relationship does not hold, and in fact when $f = f_0$ the transmitted vibration and thus the noise generated will be increased by the resilient mounting.

The natural frequency, f_0, of the support with the applied load is related to the static deflection when the load is applied. There is also a dynamic stiffness factor, k, to be considered which is due to the difference between the static stiffness of the support and the dynamic stiffness. Little is known about the values of k, but for rubber it is about 2 and for cork about 4. For a perfectly elastic material $k = 1$. f_0 is given by:

$$f_0 = \sqrt{\frac{254k}{h}} \text{ Hz (approx.)} \quad . \quad . \quad (38)$$

where h is the static deflection in millimetres.

Combining relationships (37) and (38) gives:

$$R = 20 \log \left(\frac{f^2 . h}{254k}\right) \text{ dB} \quad . \quad . \quad (39)$$

where f is the driving frequency.

Any noise being generated in a remote room will be composed of the driving frequency and possibly harmonics of it. These harmonics may be generated either by the machinery itself, or by parts of the structure. These harmonics might be louder, due to the response of the ear, than the driving frequency note. In either case, the resilient mounting should be chosen to deal with the lowest driving frequency of the machine. Then harmonics generated by the machine itself will be reduced by the amount R as given by relationship (39) plus 12 dB for every octave the harmonic is above the lowest driving frequency; harmonics generated in the structure will be attenuated by R.

For example, suppose a machine rotating at 2000 r.p.m. is generating a noise level of 60 dB in a remote room. If it is desired to reduce this noise to 20 dB then

$$40 = 20 \log \left(\frac{f^2 . h}{254k}\right).$$

$$f \text{ is } \frac{2000}{60} \text{ Hz,}$$

thus
$$\frac{h}{k} = 22 \text{ mm.}$$

If $k = 2$, then h, the static deflection under load, must be 44 mm.

IO

Criteria for Noise Control
and Sound Insulation

Noise (defined in British Standard 661:1955 as sound which is undesired by the recipient) in buildings is only important in so far as it affects people working in the buildings, and it can affect them in the following ways. It can be so loud as to cause immediate damage to the ear; it can be loud enough to cause permanent damage to the ear if the person is exposed to it for long enough; it can be loud enough to interfere with listening to speech or music; or it can be annoying.

Individuals vary in their reactions to noise and the problems must be dealt with statistically. When it is a question of damage to hearing or of interference with speech or music, while there is some variation between individuals (some people's ears seem to be more easily damaged than others'—some people speak louder than others) this variation is not great and it is possible to specify within reasonable limits what the effects of noise will be. However, when it is a question of annoyance then there is—as is well known—an almost infinite range of responses.

Nevertheless it is often essential to make some estimate of the response of a group or class of people to a particular noise environment. In recent years some criteria have been developed for this purpose and are given here. Their country of origin is mentioned because not only will there be variations in response between different groups of people in one country but also probably between similar groups of people but of different nationalities. These criteria are few and tentative; for the many noise problems for which no criterion has yet been proposed (e.g. noise in hospitals) some general working recommendations are made in Chapter 8.

For many of the criteria given here, e.g. the effect of noise on speech, it is assumed that the noise is reasonably continuous. If it is not continuous, e.g. if it is due to a passing aircraft, then obviously the criterion must also be related to how long the noise persists above the criterion being used.

CRITERION I: DAMAGE TO HEARING

Noise loud enough (about 150 dB) to cause immediate damage to hearing will not normally occur in buildings, and we will not deal with it here. Burns has given minimum sound pressure level values, in specific frequency bands, which should

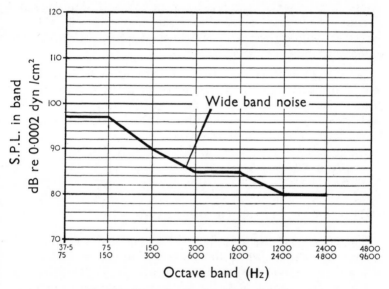

FIG. 98. Tentative Deafness Risk Criterion

not be exceeded if permanent deafness is to be avoided. These values only apply to broadband, steady noise, to which persons are exposed for the whole working day, five days per week. The values are given in Fig. 98.

The sound pressure levels may be exceeded if the exposure duration is reduced, but the relationships are only approximately known. On the other hand, it is very probable that the minimum values shown in Fig. 98 could be reduced still further if the noise is not more-or-less continuous, but instead contains one or more strong, discrete frequencies. An example would be the screech from a band-saw. But most factory noises will be of sufficiently wide band to be dealt with by the criterion shown in the Figure.

For very short exposures—of the order of one minute—much higher levels can be tolerated, but even then it is desirable not to exceed 140 dB. At these very high levels the effects on hearing seem to be independent of frequency, so the 140 dB should be taken to refer to the full-pass level.

When the noise is mainly impulsive in character the instantaneous peak levels will be much higher than the average level as indicated by an ordinary meter. No detailed information is available as to what peak levels can be tolerated, but it would of course be safe to assume that the peak levels should not exceed the values given in Fig. 98. Special apparatus is necessary to measure these peak levels.

Noise levels high enough to cause permanent deafness occur only in industry, and the procedure is obviously to measure the noise levels in octave bands at the positions that people will be and to see how they compare with the criterion. If the measured levels do exceed the safe levels one or more of the possible techniques to protect the workers should be adopted, as described in Chapter 8.

CRITERION II: SPEECH COMMUNICATION

The noise levels that interfere with speech under various conditions are known when the noise is continuous and has a reasonably continuous spectrum (after Beranek). These figures are based on average male speakers and average listeners, and there will obviously be some variation due on the one hand to those who speak louder or softer than average and on the other hand to those whose hearing is better or worse than average. This variation is of the order of ±10 dB.

The noise levels are defined in terms of the 'Speech Interference Level' (SIL), which is the average, in decibels, of the sound-pressure levels of the noise in the three octave bands 600 to 1200, 1200 to 2400 and 2400 to 4800 Hz. Table XXI shows the SIL's (in dB above 0.0002 dyn/cm^2) which 'barely permit reliable conversation at the distances and voice levels indicated'. When a woman is speaking the permissible levels should be reduced by 5 dB.

PLATE I. Sectional Model of the L.C.C. South Bank Small Concert Hall
Architect: Hubert Bennett, F.R.I.B.A.   Consultant: H. Creighton, A.R.I.B.A.

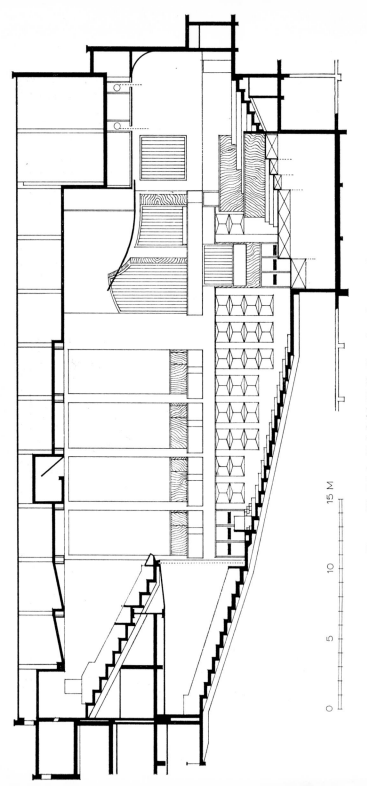

PLATE II. Main Fairfield Hall, Croydon, long section
Architects: Robert Atkinson & Partners. Consultant: H. Bagenal, F.R.I.B.A.

PLATE III. Royal Festival Hall, London
Architects: R. H. Matthew, F.R.I.B.A. and J. L. Martin, F.R.I.B.A.

PLATE IV. Lincoln Centre, Philharmonic Hall, New York
Architect: Max Abramovitz. Consultants: Bolt, Beranek & Newman

PLATE V. A B.B.C. Talks Studio

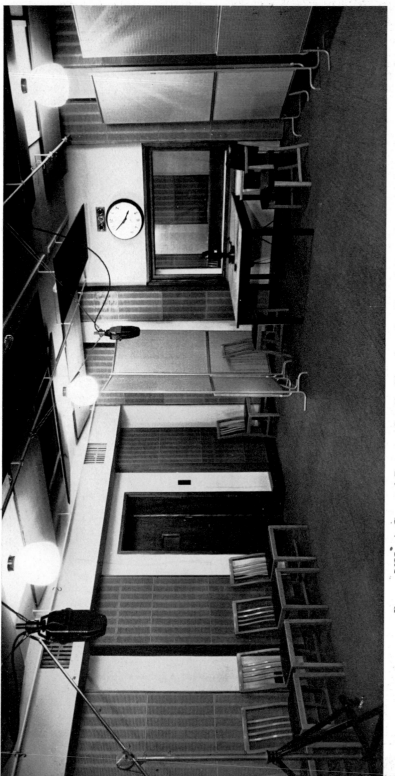

PLATE VII. A General Purpose Studio Showing the Use of Acoustic Screens

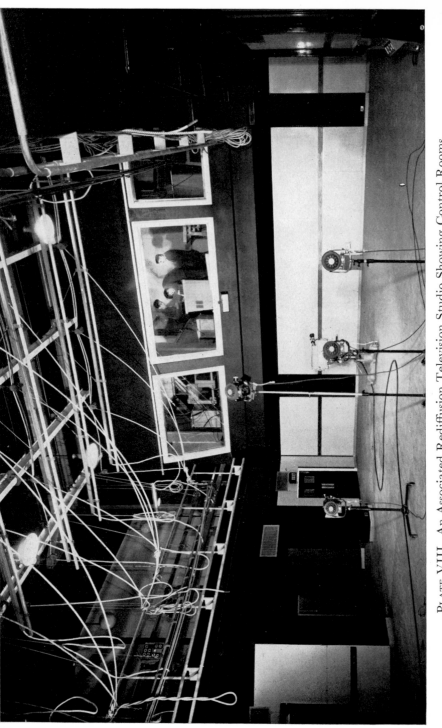

PLATE VIII. An Associated Rediffusion Television Studio Showing Control Rooms

PLATE IX. Television Announcers' Studio

PLATE X. Two Column Loudspeakers (one with the front grille removed) being te
in Salisbury Cathedral (Crown Copyright Reserved)

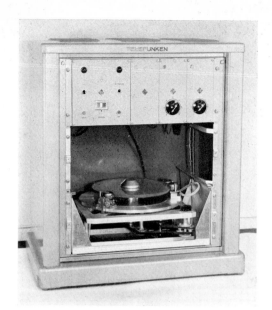

PLATE XI. Time-Delay Mechanism

TE XII. Absorbent Ceiling in the form of Deep Baffles beneath a Lay Light

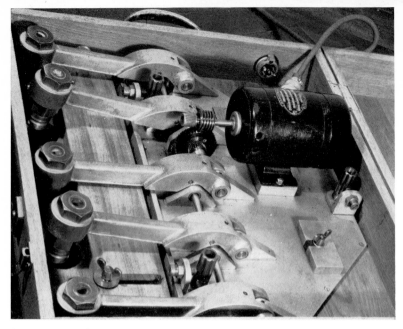

PLATE XIII. Standard Impact Machine (Crown Copyright Reserved)

PLATE XIV. Precision Sound Level Meter

TABLE XXI

SPEECH INTERFERENCE LEVELS (I.E. AVERAGE OF THE THREE
OCTAVES BETWEEN 600 AND 4800 Hz) IN dB ABOVE
0·0002 DYN/CM^2

Distance between Talker and Listener (m)	Voice Level			
	Normal	Raised	Very Loud	Shouting
0·15	71	77	83	89
0·3	65	71	77	83
0·6	59	65	71	77
1·0	55	61	67	73
1·2	53	59	65	71
1·5	51	57	63	69
1·8	49	55	61	67
3·6	43	49	55	61

The figures given in the table apply when there are no reflecting surfaces nearby and the listener and talker are facing each other. In some rooms where there are helpful reflecting surfaces, then for the larger distances between talker and listener the noise levels could be a little higher than shown.

While it is mainly noise in the three octave bands specified for the SIL's which interferes with speech, noise at lower frequencies will also interfere if it is loud enough. Fig. 99 shows the spectra corresponding to SIL's between 20 and 70; the noise at the lower frequencies should not exceed the levels shown. For example, if an SIL of 50 is required the noise in the 37 to 75 Hz octave must not exceed 83 dB; in the 75 to 150 Hz octave 72 dB, and so on.

However, while speech will be possible the low-frequency noises as specified in Fig. 99 will cause considerable strain. To avoid this strain the levels given in Fig. 100 should not be exceeded. Thus, for example, if an SIL of 50 is required the noise in the 37–75 Hz octave should not exceed 73 dB; in the 75–150 Hz octave 66 dB, and so on.

The application of the SIL criterion is straightforward. If the noise source is in the same room then obviously, from measurements of the noise level, it is possible to decide under what conditions speech will be possible. Or it is possible to decide by how much the noise must be reduced to make speech under a given condition possible. If the noise is not in the same room it will usually be a question of deciding how much sound insulation between the noise and the room will be necessary to make

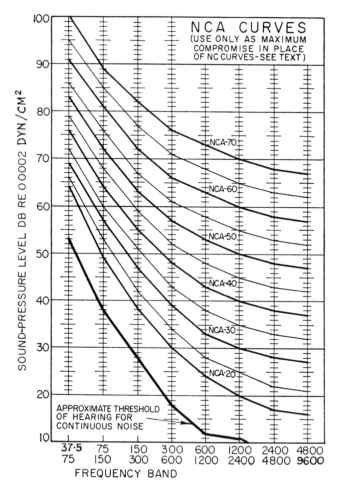

FIG. 99. Speech Interference Levels: With Strain

speech possible. An example is an office adjacent to a factory and in which speech in normal voices at 0·6 m is necessary. This calls for an SIL of 59. If some strain is tolerable then from Fig. 99 the maximum permissible noise levels in the office are as given in Table XXII. The assumed factory noise levels and thus the necessary differences are also given in the Table XXII and are plotted in Fig. 101. A suitable partition that will give this insulation can then be selected. In this case any solid, homo-

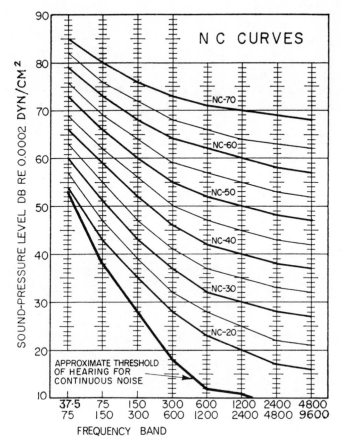

Fig. 100. Speech Interference Levels: No Strain

geneous partition weighing about 15 Kg/m² would be suitable, and its insulation graph is also plotted on Fig. 101. The fact that its insulation is more than is necessary at the higher and the lower frequencies cannot be helped; it is not yet possible to tailor an insulation graph to suit.

TABLE XXII

Octave Band (Hz)	37–75	75–150	150–300	300–600	600–1200	1200–2400	2400–4800	4800–9600
Factory Noise Levels (dB above 0·0002 dyn/cm²)	90	92	92	90	84	78	71	60
Permissible Noise Levels	90	80	71	65	62	59	57	56
Difference: i.e. insulation required in dB	0	12	21	25	22	19	14	4

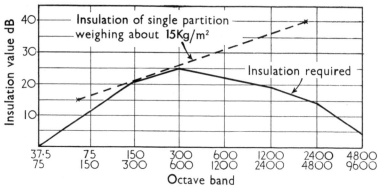

FIG. 101. Example of Use of Speech Interference Levels

A similar calculation for speech under the same conditions but without strain shows that the partition would have to be about 30 Kg/m².

A complete description of sound insulation is given in Chapter 7.

CRITERION III: OFFICES

Two criteria, to be used jointly, are proposed by Beranek (U.S.A.) for noise levels in offices. The first criterion is the Speech Interference Level (SIL) already described on p. 288. The noise rating of 'executive' offices', i.e. offices where speech is important, in terms of the SIL is given in Fig. 102. It is seen, for example, that such an office with an SIL of about 30 dB will be considered 'quiet' by its occupants, while if the SIL is about 55 dB it will be considered 'noisy'.

The second criterion is the loudness of the noise in phons, (calculated by the method due to S. S. Stevens, p. 264), and this loudness level should not exceed the SIL by more than 22 units (e.g. if the SIL is 40 dB then the loudness should not be more than 62 phons).

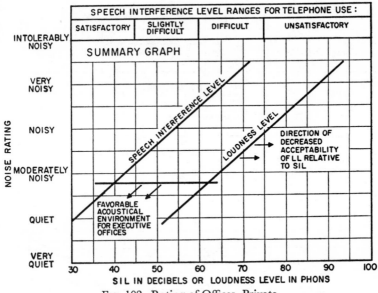

FIG. 102. Rating of Offices, Private

Fig. 100 is used for calculating the SIL's for offices where the annoyance caused by the noise, as well as the interference with speech, is important. In some circumstances it might not be too serious for the low-frequency part of the noise to be more than shown in Fig. 100; this would not cause any more interference with speech but could be more annoying. The alternative values of Fig. 99 are suggested for calculating the SIL's for these cases.

For large, open offices, such as typing pools, where noise and speech communication are not so important, the ratings of Fig. 103 apply. The effect of the noise on telephoning is shown at the top of both Figs. 102 and 103.

These criteria apply both to intruding noises and to noises made in the offices themselves. It should be noted, however, that

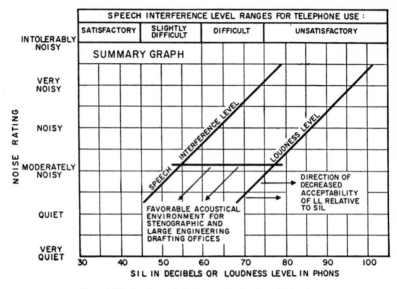

Fig. 103. Rating of Offices, Clerical and Typing

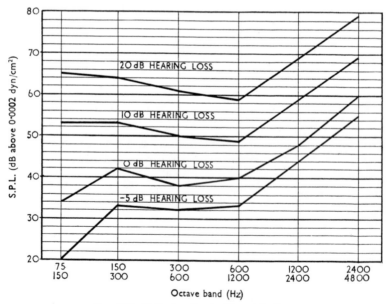

Fig. 104. Criterion for Audiometric Rooms

internal noises may not be so critical because they may be under the control of the office personnel. For example, typewriters produce an SIL of 50 to 60 dB but they can be stopped while a meeting or a telephone conversation is in progress. These criteria apply to steady noises. If the noises are intermittent, e.g. passing aircraft, then it appears that the SIL averaged over a long period will still indicate the noisiness rating of the office.

CRITERION IV: AUDIOMETRIC ROOMS

The maximum permissible noise levels in a room used for audiometry are shown in Fig. 104 (after J. R. Cox). The usual requirement is to be able to measure the threshold of hearing for people with normal hearing, and then the curve labelled 0 dB hearing loss applies. If people with more acute hearing than normal are to be tested then the − 5 dB curve applies. Similarly, if only people with some hearing loss are to be tested, then higher noise levels can be tolerated, and these levels are shown in Fig. 104, corresponding to hearing losses of 10 dB and of 20 dB.

CRITERION V: MAXIMUM NOISE LEVELS IN ROOMS USED FOR SPEECH AND MUSIC

Four criteria for permissible noise levels are given in Table XXIII (based mainly on the NC curves of Fig. 100). These levels refer of course to intruding noise and not to any internal noise in the room, e.g. due to the audience, and it is assumed that the noise is 'meaningless', i.e. is due to traffic or ventilation plant or similar sources. If the intruding noise is speech or music then, depending on the circumstances, it might need to be inaudible. Thus criterion VIII will be more appropriate.

Criterion A should be used for concert-halls where the best possible conditions are wanted. But criterion B can be used if it is not possible to get A and will still be reasonably satisfactory. Criterion B also applies to broadcasting studios, opera-houses and theatres with more than 500 seats. Criterion C applies to theatres with up to 500 seats, music rooms (in schools and similar places), classrooms, conference rooms for 50 and assembly halls. Criterion D is for cinemas, churches, court-rooms and conference rooms for 20.

TABLE XXIII

MAXIMUM PERMISSIBLE NOISE LEVELS
(dB ABOVE 0·0002 DYN/CM^2)

Octave Band (Hz)	Criteria			
	A	B	C	D
37–75	53	54	57	60
75–150	38	43	47	51
150–300	28	35	39	43
300–600	18	28	32	37
600–1200	12	23	28	32
1200–2400	11	20	25	30
2400–4800	10	17	22	28
4800–9600	22	22	22	27

CRITERION VI: AIR-BORNE SOUND INSULATION

The air-borne sound insulation of all walls and floors varies with frequency and in noise problems it would often be desirable to specify the necessary insulation as a function of frequency. However, while this can sometimes be done there are many problems where it is not done, either because there is not sufficient information to justify it or because such accuracy is not necessary.

Thus, as a rough working rule the insulation of a construction is often given as a single figure—the average for the frequency range 100 to 3150 Hz. A slight increase in the accuracy of this rule can be got by making an approximate allowance for the 'slope' of the insulation (i.e. the average increase in insulation per octave).

Most normal partitions used in buildings have 'slopes' of between 4 and 10 dB per octave and it has been found that a partition of slope of 10 dB per octave is 3 dB less effective in reducing the loudness (but not necessarily the annoyance) of male speech than a partition of the same average insulation but with a slope of 5 dB per octave. Similarly, a 10-dB slope partition is 2·5 dB less effective against light music than a 5-dB slope partition.

Thus for example a wall of average insulation of 40 dB and

with a slope of 5 dB per octave would reduce the loudness of male speech as much as a wall of 43 dB average insulation and with a slope of 10 dB per octave.

CRITERION VII: SOUND INSULATION BETWEEN DWELLINGS

Insulation between dwellings is one case where the single-figure average for air-borne sound insulation is not sufficiently precise. Instead, the insulation is specified over the whole frequency range and the insulation of a construction should not come below it at any part of the frequency range by more than a specified amount. The reason for this method is that, once insulation is adequate over one part of the frequency range, further insulation over that part is of no further advantage and will not affect any deficiency in insulation over another part of the range, although it would increase—misleadingly—the average figure.

Several countries now specify sound insulation in this way; we give here two examples.

The first refers to new dwellings in the United Kingdom under the regulations mentioned on p. 191.

For walls between houses the required air-borne sound insulation (which is specifically stated only in the Scottish regulations) is shown in Fig. 105. Any deviation in the unfavourable direction should not exceed 1 dB when averaged over the whole frequency range.

For party walls and party floors between flats (under Scottish Regulations) the requirements are for the Grade I insulation shown in Fig. 106. Grade I represents the highest insulation that is practicable for flats and with it the noise from neighbours causes only minor disturbance to most tenants; it is no more nuisance than other minor disadvantages of living in flats. Grade II is a lower value of insulation and with it many of the tenants consider the noise from their neighbours to be the worst factor about living in flats, but even so at least half of the tenants will consider themselves not seriously disturbed by the noise. If the insulation is less than Grade II then the number of tenants seriously disturbed will increase, until a level of insulation as low as 8 dB worse than Grade II is reached when strong reactions, i.e. deputations, are probable.

To satisfy the particular grade the insulation of the construction must be such that any deviations in the unfavourable direction should not exceed 1 dB when averaged over the whole frequency range. When the mean unfavourable divergence below Grade II does exceed 1 dB, then this mean divergence is given as so many dB worse than Grade II.

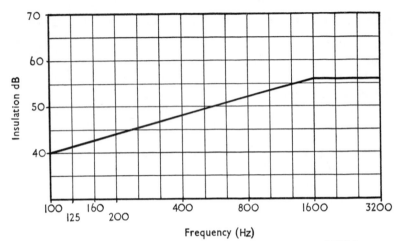

FIG. 105. Air-borne Sound Insulation between Houses (U.K.)

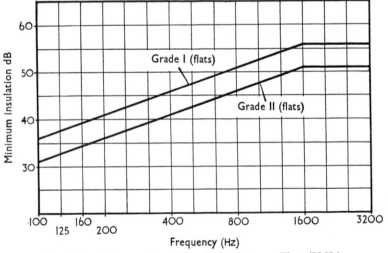

FIG. 106. Air-borne Sound Insulation between Flats (U.K.)

For example, Fig. 107 shows the air-borne insulation of a concrete floor with a floating raft plotted for comparison with the grade curves. The shaded area shows where this floor falls below Grade I, but the mean divergence is less than 1 dB, so this floor would be classified as Grade I for air-borne sound. (The mean divergence is obtained from the sum of the 1-dB divergence at 100 Hz, the 2 dB at 125 Hz, the 2·5 dB at 160 Hz, the 1 dB at 200 Hz and the 1 dB at 250 Hz, i.e. a total of 7·5 dB, which, to get the average divergence over the whole frequency range, has to be divided by 16, being the number of separate frequencies in the range 100 to 3150 Hz.)

For impact sound insulation the Scottish regulation specifies the Grade I values (measured without lino) shown in Fig. 108.

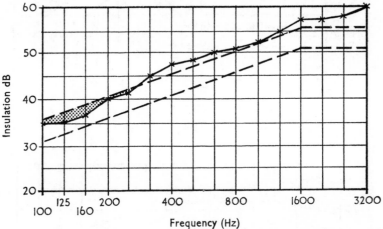

FIG. 107. Example of Air-borne Sound Insulation Grading

It should be remembered that impact sound insulation (see Chapter 9) is given in terms of the sound-pressure levels in the receiving room when the standard impacts machine is operated on the floor above. Thus the higher the levels the less the insulation. The values shown in Fig. 108 are the sound-pressure levels (in dB relative to 0·0002 dyn/cm^2) in octave bands which must not be exceeded in the receiving room. The same conditions for divergencies in the unfavourable direction apply as for air-borne sound insulation.

The impact sound insulation of a wood-joist floor with some sound-insulation treatment is shown in Fig. 109 plotted against the grades. It is seen that the insulation comes below Grade II over the frequency range 100 to 400 Hz and the mean divergence is 2 dB. This floor would thus be classified as 2 dB worse than Grade II for impact sound insulation.

To satisfy a particular grade a floor construction must be satisfactory for both air-borne and impact sound insulation.

The second example is the German provisional standard DIN52211. This deals with both laboratory and field measure-

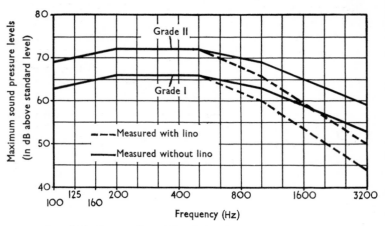

FIG. 108. Impact Sound Insulation between Flats (U.K.)

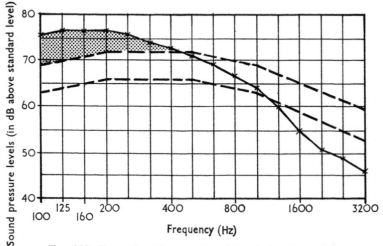

FIG. 109. Example of Impact Sound Insulation Grading

ments but here we give only the field standards. They are shown in Figs. 110 and 111. The air-borne measurements are corrected by relationship (11) and the impact insulation by relationship (23) (see Chapter 9) and the mean divergence permitted in the unfavourable direction is 2 dB. (Any deficiencies at the two extreme frequencies of 100 and 3150 Hz are counted as half their numerical value when calculating the mean divergence.)

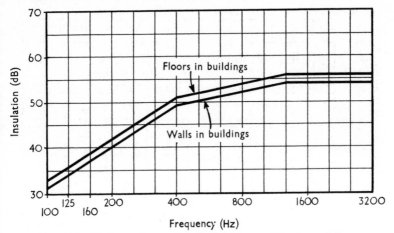

FIG. 110. Air-borne Sound Insulation between Houses and Flats (Germany)

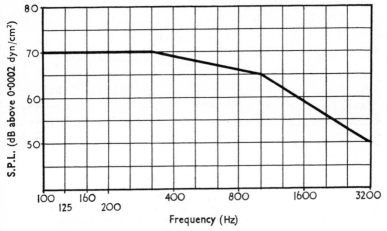

FIG. 111. Impact Sound Insulation between Flats (Germany)

CRITERION VIII: INAUDIBILITY

Fig. 112 shows the approximate threshold of hearing for continuous noises. For example, a noise getting into a room must be reduced to 53 dB in the 37–75 Hz octave, to 38 dB in the 75–150 Hz octave and so on, if it is to be reduced to inaudibility. This applies only when there is no other noise in the receiving room. Obviously if there is some other noise then the intruding noise will not have to be reduced so much.

The masking of one noise by another is a complicated phenomenon and we will here only give a rough working guide. It is assumed that the masking noise has the spectrum typical of many traffic and industrial (e.g. ventilation plant) noises, which is fairly continuous and falling by about 3 dB per octave. Then

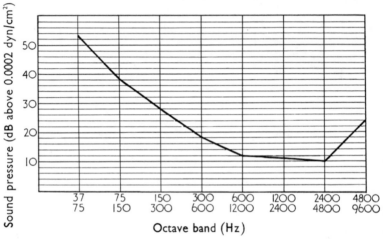

FIG. 112. Threshold of Audibility for Continuous Noise

if the level of the masking noise is $(40 + x)$ dB in the 300–600 Hz octave, then the masking levels (i.e. the amounts by which the thresholds of audibility are raised) are as shown in Table XXIV.

TABLE XXIV

Octave Band (Hz)	Rise in Threshold (dB)
37–75	$2 + x$
75–150	$8 + x$
150–300	$14 + x$
300–600	$20 + x$
600–1200	$24 + x$
1200–2400	$25 + x$
2400–4800	$27 + x$
4800–9600	$11 + x$

For example, suppose a noise in one room of level 80 dB in the 150–300 Hz octave has to be reduced to inaudibility in another room. If there were no masking noise in the second room then this would have to be reduced by 52 dB to 28 dB (from Fig. 112). In the second room suppose the masking noise level in the 300–600 Hz band (e.g. due to traffic) is 45 dB. Then $x = 5$ dB, and

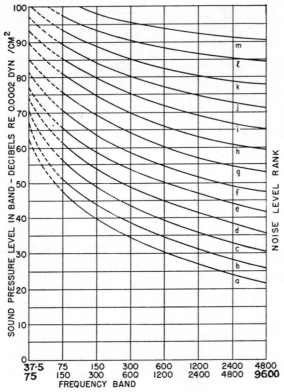

FIG. 113. Noise Level Rank

the rise in the threshold level in the 150–300 Hz band is thus 19 dB, i.e. the threshold is now 47 dB. The reduction required is now $80 - 47 = 33$ dB.

CRITERION IX: NOISE OUT-OF-DOORS

One general criterion (after Stevens *et al.*, U.S.A.) is given here for estimating the annoyance likely to be caused to a neighbourhood by the introduction of a new noise source, e.g. a factory, into it. Alternatively, the criterion may be used to decide whether or not a specific complaint about an existing noise is reasonable or not. It should be noted that the noise is actually measured out-of-doors, although the annoyance caused by it refers of course to people living in houses. The procedure is as follows.

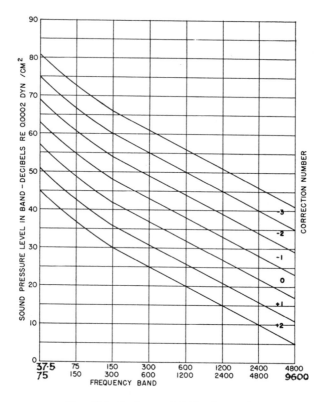

Fig. 114. Background Noise Correction

The spectrum of the intruding noise, as measured or as predicted at the appropriate position, is plotted on Fig. 113. The highest zone into which the spectrum protrudes decides the 'noise level rank' ranging from 'a' to 'm'. Four corrections are then applied. The first depends on the background noise, i.e. the noise present at the site in the absence of the specific intruding noise. The spectrum of the background noise is plotted on Fig. 114 and the zone in which the major portion of the spectrum lies gives the correction number, of between $+2$ to -3. If the background noise level is not known Table XXV can be taken as a guide to typical areas during the day-time.

TABLE XXV

CORRECTION NUMBERS FOR BACKGROUND NOISE

Area	Correction Number
Very quiet suburban	$+1$
Suburban	0
Residential urban	-1
Urban near some industry	-2
Area of heavy industry	-3

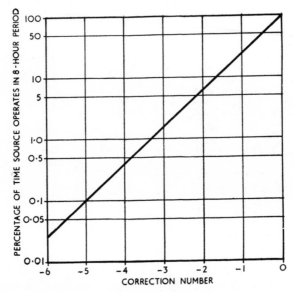

FIG. 115. Correction for Repetitiveness

The second correction depends on the time factors. If the noise occurs during the day-time only, a correction of -1 is applied. If it is in the winter only, the correction is -1. Corrections between 0 and -6 are applied depending on what percentage of time during an eight-hour period the noise operates, as shown in Fig. 115. If the noise operates only one or two days a week, then an additional correction of -1 is required.

The third correction depends on the nature of the noise. If it has a reasonably continuous spectrum then no correction is applied. If on the other hand it consists mainly of a few more or less pure tones then the correction is $+1$. Also, if it is an impulsive noise such as is made by drop-hammers, then the correction is $+1$.

The fourth correction depends on whether the residents have had some previous experience of the noise, or of some similar noise. If they have had previous experience then the correction is -1.

All these corrections are summarised in Table XXVI.

TABLE XXVI

Influencing Factor	Correction
1. Background Noise (See Fig. 114 and Table XXV)	$+2$ to -3
2. Time Factors	
Day-time only	-1
Repetitiveness (from Fig. 115)	0 to -6
One or two days only per week	-1
Winter only	-1
3. Noise Character	
Pure-tone components	$+1$
Impulsive	$+1$
4. Some previous exposure	-1

All the correction numbers are added together, and the sum applied to the original noise level rank (from Fig. 113) to get what is called the Composite Noise Rating. For example, if the sum of the corrections comes to $+3$ and the original noise level rank was d, then the Composite Noise Rating would become G. Finally, from Fig. 116 the Composite Noise Rating will show the average expected response, together with the range of response that might be expected from average communities.

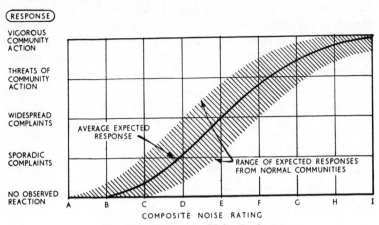

FIG. 116. Composite Noise Rating

CRITERION X: ZONING REGULATIONS

To ensure that noise emanating from factories and the like should not interfere unduly with nearby residential buildings it is possible to specify the maximum noise levels that should be permitted at the boundary between the two types of buildings. Such specifications are rare at the present time but they will probably become more common. Two examples will be given here.

The first has been specified tentatively for Chicago (1956). It states that the noise levels at the boundary between industrial and residential areas should not exceed:

Octave Band (Hz)	37– 75	75– 150	150– 300	300– 600	600– 1200	1200– 2400	2400– 4800	4800– 9600
Sound-pressure Level (dB above 0·0002 dyn/cm²)	72	67	59	52	46	40	34	32

For a boundary between a commercial and an industrial area the proposed figures are 7 dB higher at all frequencies.

These permissible levels were obtained empirically by examinations of actual acceptable conditions in Chicago.

The second example, also from the U.S.A., is in the town of Stony Point, New York, where a proposed (1954) zoning ordinance lays down sound-pressure levels in octave bands that may not be exceeded at the boundary between any industrial area and any residential or commercial area. These levels are:

Maximum permissible sound-pressure levels at specified points of measurement for noise radiated continuously from a facility at night-time:

Frequency Band (Hz)	20–75	75–150	150–300	300–600	600–1200	1200–2400	2400–4800	4800–10,000
Sound-pressure Level in dB above 0·0002 dyn/cm²	69	54	47	41	38	38	38	38

If the noise is not smooth and continuous and is not radiated at night-time one or more of the following corrections shall be added to or subtracted from each of the decibel levels given above:

Type of Operation or Character of Noise	*Correction in Decibels*
Day-time operation only	+5
Noise source operates less than 20% of the time	+5*
Noise source operates less than 5% of the time	+10*
Noise source operates less than 1% of the time	+15*
Noise of impulsive character (hammering, etc.)	−5
Noise of periodic character (hum, screech, etc.)	−5

* Apply one of these corrections only.

Absorption Coefficients

The following table of absorption coefficients is divided into four groups: common building materials (1 to 22), common absorbent materials of non-proprietary kinds (23 to 42), room contents (43 to 49), and proprietary absorbents (50 to 69). Coefficients are given for the three representative frequencies 125, 500 and 2000 Hz at which calculations are commonly made and also for a number of materials at some or all of the frequencies 62, 250, 1000 and 4000 Hz to enable calculations at every octave over a wider range to be made for studio design purposes. In all of the groups except the proprietary materials the values given are those which have been found in practice to be most applicable to average room and auditorium conditions, rather than values based on an isolated test measurement. It must be borne in mind that sound absorption is not an intrinsic property of a material alone. Factors such as thickness, method of mounting and decorative treatment will influence actual absorption, as will the nature (solidity and weight, for example) of the structures in which they are built, particularly at the lowest sound frequencies.

The values for proprietary absorbents are those published by the manufacturers of these materials, and only those which are results of tests by the National Physical Laboratory of Great Britain and the Technical Physics Dept. (T.N.O.) of the Netherlands (which are recognised authorities) have been included.

The values quoted under the heading L.R.C. are the loudness reduction coefficients which give an indication of the performance of the material as a noise-reducing treatment (see Chapter 8, p. 226).

| | \
 \
 Frequency Hz | | | | | | |
|---|---|---|---|---|---|---|---|---|
| | 62 | 125 | 250 | 500 | 1000 | 2000 | 4000 | L.R.C. |
| COMMON BUILDING MATERIALS | | | | | | | | |
| 1. Boarded roof; underside of pitched slate or tile covering | | 0·15 | | 0·1 | | 0·1 | | |
| 2. Boarding ('match') about 19 mm thick over air-space against solid wall | | 0·3 | | 0·1 | | 0·1 | | |
| 3. Brickwork, plain or painted | 0·05 | 0·05 | 0·04 | 0·02 | 0·04 | 0·05 | 0·05 | |
| 4. Clinker ('breeze') concrete unplastered | 0·1 | 0·2 | 0·3 | 0·6 | 0·6 | 0·5 | 0·5 | |
| 5. Concrete, constructional or tooled stone or granolithic | 0·05 | 0·02 | 0·02 | 0·02 | 0·04 | 0·05 | 0·05 | |
| 6. Cork tiles (thin), wood blocks, linoleum or rubber flooring on solid floor (or wall) | 0·05 | 0·02 | 0·04 | 0·05 | 0·05 | 0·1 | 0·05 | |
| 7. Cork tiles 25 mm thick on solid backing | | 0·05 | 0·1 | 0·2 | 0·55 | 0·6 | 0·55 | 0·5 |
| 8. Fibreboard (normal soft) 12 mm thick, mounted against solid backing—unpainted | 0·05 | 0·05 | 0·1 | 0·15 | 0·25 | 0·3 | 0·3 | |
| 9. Ditto, painted | 0·05 | 0·05 | 0·1 | 0·1 | 0·1 | 0·1 | 0·15 | |
| 10. Fibreboard (normal soft) 12 mm thick mounted over 25 mm air-space on battens against solid backing—unpainted | | 0·3 | | 0·3 | | 0·3 | | |
| 11. Ditto, painted | | 0·3 | | 0·15 | | 0·1 | | |
| 12. Floor tiles (hard) or 'composition' floor | | 0·03 | | 0·03 | | 0·05 | | |
| 13. Glass; windows glazed with up to 4 mm glass | | 0·3 | | 0·1 | | 0·05 | | |
| 14. Glass, 6 mm plate windows in large sheets | | 0·1 | | 0·04 | | 0·02 | | |
| 15. Glass used as a wall finish (e.g. 'Vitrolite') or glazed tile or polished marble | | 0·01 | | 0·01 | | 0·02 | | |
| Granolithic floor—see 5 | | | | | | | | |
| Lath and plaster—see 17 | | | | | | | | |
| Linoleum—see 6 | | | | | | | | |
| Marble—see 15 | | | | | | | | |
| Match-boarding—see 2 | | | | | | | | |
| 16. Plaster, lime or gypsum on solid backing | 0·05 | 0·03 | 0·03 | 0·02 | 0·03 | 0·04 | 0·05 | |
| 17. Plaster, lime or gypsum on lath, over air-space against solid backing or on joists or studs including plasterboard | 0·1 | 0·3 | 0·15 | 0·1 | 0·05 | 0·04 | 0·05 | |
| 18. Plaster or plasterboard suspended ceiling with large air-space above | | 0·2 | | 0·1 | | 0·04 | | |
| 19. Plywood or hardboard panels mounted over air-space against solid backing | | 0·3 | | 0·15 | | 0·1 | | |

	Frequency Hz							
	62	125	250	500	1000	2000	4000	L.R.C.
COMMON BUILDING MATERIALS—*continued*								
20. Ditto with porous absorbent in air-space		0·4		0·15		0·1		
Rubber flooring—see 6								
Stone, polished—see 15								
21. Water—as in swimming-baths		0·01		0·01		0·02		
Windows—see 13 and 14								
Wood-block floor—see 6								
22. Wood boards on joists or battens	0·1	0·15	0·2	0·1	0·1	0·1	0·1	
COMMON ABSORBENT MATERIALS (NON-PROPRIETARY)								
23. Asbestos spray, 25 mm on solid backing—unpainted		0·15		0·5		0·7		
24. Carpet—thin, such as hair cord over thin felt on concrete floor	0·05	0·1	0·15	0·25	0·3	0·3	0·3	0·3
25. Ditto on wood-board floor	0·15	0·2	0·25	0·3	0·3	0·3	0·3	0·3
26. Carpet, pile over thick felt on concrete floor	0·05	0·07	0·25	0·5	0·5	0·6	0·65	0·55
27. Curtain—medium or similar fabric, straight against solid backing	0·05	0·05	0·1	0·15	0·2	0·25	0·3	0·2
28. Curtain medium fabric hung in folds against solid backing		0·05		0·35		0·5		
29. Curtains (dividing), double, canvas	0·03	0·03	0·04	0·1	0·15	0·2	0·15	
30. Felt—hair, 25 mm thick with perforated membrane (viz. muslin) against solid backing		0·1		0·7		0·8		
Mineral or glass wool, 80–190 Kg/m³ density, 25 mm thick blanket or semi-rigid slabs against solid backing:								
31. With no covering, or very porous (scrim or open-weave fabric) or open metal mesh covering	0·08	0·15	0·35	0·7	0·85	0·9	0·9	0·85
32. With 5% perforated hardboard covering	0·05	0·1	0·35	0·85	0·85	0·35	0·15	0·55
33. With 10% perforated or 20% slotted hardboard covering	0·05	0·15	0·3	0·75	0·85	0·75	0·4	0·7
Mineral or glass wool, 80–190 Kg/m³ density, 50 mm thick blanket or mattress mounted over 25 mm air-space against solid backing:								
34. No covering or with very porous (scrim or open-weave fabric) or open metal mesh covering	0·15	0·35	0·7	0·9	0·9	0·95	0·9	0·9
35. Ditto with 10% perforated or 20% slotted hardboard covering	0·15	0·4	0·8	0·9	0·85	0·75	0·4	0·7

	Frequency Hz							
	62	125	250	500	1000	2000	4000	L.R.C.
COMMON ABSORBENT MATERIALS (NON-PROPRIETARY)–*continued*								
36. Panel (about 5 Kg/m²) of 3 mm hardboard with bitumen roofing felt stuck to back mounted over 50 mm air-space against solid backing	0·5	0·9	0·45	0·25	0·15	0·1	0·1	
37. Panel (about 4 Kg/m²) of two layers bitumen roofing felt mounted over 250 mm air-space against solid backing	0·9	0·5	0·3	0·2	0·1	0·1	0·1	
38. Polystyrene (expanded) board 25 mm thick spaced 50 mm from solid backing		0·1	0·25	0·55	0·2	0·1	0·15	0·25
39. Polyurethane flexible foam 50 mm thick on solid backing		0·25	0·5	0·85	0·95	0·9	0·9	0·9
40. Wood-wool slabs 25 mm thick mounted solidly—unplastered		0·1		0·4		0·6		
41. Ditto mounted 25 mm from solid backing		0·15		0·6		0·6		
42. Ditto, plastered and with mineral wool in cavity		0·5		0·2		0·1		
ROOM CONTENTS								
43. Air. (*x*) (per cu. m)	nil	nil	nil	nil	0·003	0·007	0·02	
44. Audience seated in fully upholstered seats (per person)	0·15	0·18	0·4	0·46	0·46	0·51	0·46	
45. Audience seated in wood or padded seat (per person)		0·16		0·4		0·44	0·4	
46. Seats (unoccupied), fully upholstered (per seat)		0·12		0·28		0·32	0·37	
47. Seats (unoccupied), wood or padded (per seat)		0·08		0·15		0·18	0·2	
48. Orchestral player with instrument (average)	0·18	0·37	0·8	1·1	1·3	1·2	1·1	
49. Rostrum (portable wood) per m² of surface	0·6	0·4	0·1	nil	nil	nil	nil	
ABSORBENT MATERIALS, PROPRIETARY								
50. 'Burgess' metal perforated tile (type C) against solid backing		0·1	0·3	0·6	0·75	0·8	0·8	0·75
51. 'Echostop' plaster perforated tile over 125 mm air-space		0·45	0·7	0·8	0·8	0·65	0·45	0·7
52. Fibreglass 19 mm plastic filmed acoustic tiles spaced 50 mm from solid backing. (Film 0·038 mm stretched across tiles and stuck at edges only)		0·3	0·45	0·7	0·75	0·85	0·75	0·75
53. 'Frenger' metal perforated (heated) panel with 19 mm bitumen-bonded glass wool behind, over air-space		0·2	0·45	0·65	0·45	0·35	0·25	0·4

				Frequency Hz				
	62	125	250	500	1000	2000	4000	L.R.C.

ABSORBENT MATERIALS,
PROPRIETARY—*continued*

	62	125	250	500	1000	2000	4000	L.R.C.
54. 'Gypklith' wood-wool tile, 25 mm thick over 25 mm air-space		0·25	0·45	0·9	0·7	0·55	0·75	0·7
55. 'Gyproc' perforated plasterboard over 25 mm scrim-covered rock-wool		0·15	0·7	0·9	0·7	0·45	0·3	0·6
56. Ditto over 50 mm glass-wool		0·4	0·75	0·85	0·55	0·45	0·3	0·55
57. Ditto over 25 mm air-space (empty)		0·1	0·2	0·4	0·3	0·15	0·2	
58. 'Gyproc' slotted plasterboard tile over 25 mm bitumen-bonded glass-wool		0·15	0·5	0·8	0·6	0·25	0·3	0·5
59. 'Paxfelt' asbestos felt 25 mm thick over 25 mm air-space			0·5	0·55	0·65	0·7	0·75	0·65
60. 'Paxtiles' asbestos tiles 25 mm thick over 25 mm air-space			0·55	0·75	0·85	0·8		
61. 'Perfonit' wood fibre perforated tile 19 mm thick over 25 mm air-space		0·2	0·5	0·7	0·85	0·75	0·65	0·75
62. 'Tentest' Rabbit-Warren perforated hardboard tile with grooved fibre backing 25 mm mounted over 25 mm air-space		0·15	0·5	0·6	0·8	0·75	0·25	0·6
63. 'Thermacoust' wood-wool slab 50 mm thick against solid backing		0·2	0·3	0·8	0·75	0·75	0·75	0·75
64. 'Treetex', 'Decorac' slotted wood-fibre tile 25 mm thick		0·15	0·65	0·75	1·00	0·95	0·7	0·85
65. 'Treetex', 'Slotac' grooved wood-fibre tile 19 mm thick		0·15	0·4	0·55	0·7	0·8	0·7	0·7
66. 'Treetex', 'Treeperac' perforated wood-fibre tile 19 mm thick		0·2	0·55	0·65	0·9	0·8	0·55	0·7
67. 'Unitex' perforated wood-fibre tile 12 mm thick		0·2	0·55	0·6	0·6	0·65	0·8	0·65
68. 'Unitex' perforated wood-fibre tile 19 mm thick		0·25	0·65	0·65	0·7	0·8	0·75	0·7
69. 'W. Cullum' Acoustic Felt, covered with painted and pin-hole perforated muslin—solid backing			0·35	0·75	0·65	0·7	0·65	0·75

APPENDIX B

Weights of Various Common Building Materials

The following are conservative estimates of the weights of various common building materials designed for use in assessing sound-insulation properties. For this reason the values in some cases will be found to be less than the Imperial measure equivalents quoted in B.S. 648 which are designed as information for calculating dead loads and therefore provide a margin of safety. All weights are in kilograms per square metre.

Item	Material	Nominal Thickness	Weight
1.	Asphalt—rock	25 mm	53·5
2.	Blocks—walling		
	Solid:		
	Clinker concrete	per cm	11·5
	'Thermalite'	,,	7·9
	Hollow:		
	Clay	,,	7·7
3.	Boards and slabs:		
	Asbestos 'insulating'	12 mm	8·5
	Blockboard	25 mm	12·2
	Chipboard	25 mm	14·6
	Hardboard:		
	Medium	3 mm	3·4
	Super	3 mm	4·9
	Plasterboard	10 mm	8·3
	,, plastered with skim coat (3 mm)		13·2
	Plywood	1 mm	0·49
	Straw ('Stramit')	50 mm	18·5
	Wood-fibre 'insulating'	12 mm	3·4
	Wood wool	25 mm	9·7
4.	Brickwork (as laid):		
	'Common' such as London stock, Fletton or sand lime	110 mm	204
	Diatomaceous earth (flue liners)	,,	78
	Engineering, dense	,,	254

Item	Material	Nominal Thickness	Weight
5.	Concrete (poured):		
	Brick aggregate	1 cm	19·2
	Clinker aggregate	1 cm	13·4
	Sand, gravel or crushed stone aggregate (reinforced)	1 cm	23
6.	Felt—bituminous roofing	3 ply	1·9
7.	Glass	6 mm	14·6
8.	Linoleum	4·5 mm	2·9
9.	Plaster, gypsum or lime	12 mm	19·5
10.	Plaster and lath (wood or metal) 3-coat		29
11.	Rubber	3 mm	4·9
12.	Sheets:		
	Aluminium	22 s.w.g.	1·9
	Asbestos and steel ('Durasteel' fireproof)	6 mm	22
	Asbestos Cement	6 mm	9·7
	Copper	24 s.w.g.	4·9
	Steel	16 s.w.g. (1·5 mm)	12·2
	Zinc	20 s.w.g.	6·3
13.	Slate—sawn slab	25 mm	72
14.	Stone (average)	,,	51
15.	Timber:		
	Seasoned softwoods	,,	12
	Common hardwoods	,,	18

APPENDIX C

Sound Reduction Indices

As explained in Chapter 9, the sound-reduction index of a wall or floor can only be measured when indirect transmission is negligible, as happens in the laboratory for all constructions but in the field only for constructions of less than about 40 dB average insulation. There are several publications from various laboratories giving the true sound-reduction indices of many constructions, but for practical purposes some estimate of the indices is required, including the indirect transmission. This has been done in this table, but it is clear that the values given will depend to some extent on the surrounding structure. The figures given here, then, are fairly typical of these walls and floors when used with traditional construction, and when the area is of the order of 18 to 47 m². It is not possible to say what the effect of non-traditional construction will be, but if the area is greater than about 47 m² it is probable that these indices will be rather higher.

Type of Partition	100	125	160	200	250	315
SOLID WALLS						
110 mm Brick (plastered)	31	34	35	36	36	37
150 mm dense Concrete	32	29	38	37	39	39
220 mm Brick (plastered)	41	41	43	43	45	45
340 mm Brick (plastered)	42	44	46	45	43	46
WOOD JOIST FLOORS						
T. & G. boards, plasterboard ceiling with skim coat	14	18	21	23	25	26
As above but with boards 'floating' on glass wool	22	25	33	33	33	38
Boards nailed and not 'floating', 76 mm rock-wool direct on ceiling	29	29	31	33	34	35
Boards 'floating' on glass-wool and 76 mm rock-wool direct on ceiling	23	27	28	35	35	41
Boards 'floating' on glass-wool and 50 mm sand direct on ceiling of 3-coat plaster on metal lath	35	36	40	41	42	43
CONCRETE FLOORS						
Solid, 126 mm thick	32	35	36	36	36	38
As above but with 'floating' concrete screed	38	38	40	42	43	45
WINDOWS						
Single: 4 mm glass, normally closed	16	17	15	18	21	21
Double: 4 mm glass, 200 mm air-space with absorbent in the reveals; tightly sealed	26	30	29	33	35	39

INDICES OF COMMON WALLS AND
ETC.

Frequency (Hz)										Average (100–3150 Hz)
400	500	630	800	1000	1250	1600	2000	2500	3150	
38	41	45	50	51	53	55	58	59	60	45 dB
42	45	46	51	52	53	57	60	63	67	47 dB
47	48	52	54	56	56	57	58	59	60	50 dB
48	49	54	54	57	58	62	63	64	61	52 dB
34	37	41	39	39	42	45	45	43	45	34 dB
36	38	41	42	45	50	54	56	60	61	42 dB
37	39	39	40	41	45	48	50	50	50	39 dB
41	44	45	47	48	49	53	56	58	59	43 dB
46	47	49	51	52	54	57	60	62	63	49 dB
40	41	44	45	49	52	55	58	59	62	45 dB
47	48	51	52	54	56	59	61	60	63	50 dB
24	25	25	27	26	22	21	23	25	26	22 dB
41	43	43	44	46	47	46	47	43	37	39 dB

APPENDIX D

Octave Analyses of Some Common Noises

Noise	Distance, m	Sound-pressure Levels (Octave Bands)								Remarks
		37–75 Hz	75–150 Hz	150–300 Hz	300–600 Hz	600–1200 Hz	1200–2400 Hz	2400–4800 Hz	4800–9600 Hz	
Large (4 engines) jet air-liner	38	112	121	123	124	123	120	117	109	Maximum values when passing overhead at take-off power. No mufflers.
Single-engined jet fighter	38	102	114	116	116	117	115	111	102	Maximum values when passing overhead at take-off power.
Large (4 engines) piston-engined air-liner	38	111	117	114	108	107	108	106	97	Maximum values when passing overhead at take-off power.
Electric trains over steel bridge	6	94	93	99	99	95	84	81	73	
Curb-side, main road in London at rush hour	5 (average)	78	81	81	79	72	67	63	55	
Electric trains	30	77	77	76	74	73	67	59	54	
Pneumatic drills	38	75	72	72	66	69	71	67	65	
Riveting on large (6 m by 5 m) steel plate	2	88	96	105	106	111	109	113	110	In open air.
Nylon factory	Reverberant sound	87	86	92	93	97	97	96	87	Up-twisting process.
Weaving shed	Reverberant sound	78	71	77	81	86	86	84	78	Average levels. Peak values up to 20 dB higher.
Canteen (hard ceiling)	Reverberant sound	52	54	59	67	67	61	55	49	Ten typewriters, one tele-type machine
Typing office with acoustic ceiling	Reverberant sound	68	64	60	56	55	55	53	50	Average values.
Male speech	1	52	55	59	66	65	60	52	40	

APPENDIX E

Decibel Table

Intensity Ratio	Pressure Ratio	←dB→	Pressure Ratio	Intensity Ratio
1	1	0	1	1
0·80	0·89	1	1·1	1·3
0·63	0·79	2	1·3	1·6
0·50	0·71	3	1·4	2
0·40	0·63	4	1·6	2·5
0·32	0·56	5	1·8	3·2
0·25	0·50	6	2·0	4·0
0·20	0·45	7	2·2	5·0
0·16	0·40	8	2·5	6·3
0·13	0·35	9	2·8	7·9
0·10	0·32	10	3·2	10
0·08	0·28	11	3·5	13
0·063	0·25	12	4·0	16
0·050	0·22	13	4·5	20
0·040	0·20	14	5·0	25
0·032	0·18	15	5·6	32
0·025	0·16	16	6·3	40
0·020	0·14	17	7·1	50
0·016	0·13	18	7·9	63
0·013	0·11	19	8·9	79
0·010	0·10	20	10	100
10^{-1}	$3·2 \times 10^{-1}$	10	3·2	10
10^{-2}	10^{-1}	20	10	10^2
10^{-3}	$3·2 \times 10^{-2}$	30	$3·2 \times 10$	10^3
10^{-4}	10^{-2}	40	10^2	10^4
10^{-5}	$3·2 \times 10^{-3}$	50	$3·2 \times 10^2$	10^5
10^{-6}	10^{-3}	60	10^3	10^6
10^{-7}	$3·2 \times 10^{-4}$	70	$3·2 \times 10^3$	10^7
10^{-8}	10^{-4}	80	10^4	10^8
10^{-9}	$3·2 \times 10^{-5}$	90	$3·2 \times 10^4$	10^9
10^{-10}	10^{-5}	100	10^5	10^{10}

Example: To find the dB corresponding to a pressure ratio of 25.

$$\text{Ratio of } 25 = 2·5 \times 10$$
$$\text{In dB} = +8+20 \text{ dB}$$
$$= \textbf{+28 dB.}$$

Example: To find the intensity ratio corresponding to a dB ratio of -43.

$$-43 \text{ dB} = -40 \text{ dB} - 3 \text{ dB}$$
$$= 10^{-4} \times 0 \cdot 50$$
$$= 5 \times 10^{-5}.$$

APPENDIX F

Logarithms (to the base 10)

Number	Log	Number	Log	Number	Log	Number	Log
10	00	33	52	56	75	79	90
11	04	34	53	57	76	80	90
12	08	35	54	58	76	81	91
13	11	36	56	59	77	82	92
14	15	37	57	60	78	83	92
15	18	38	58	61	79	84	92
16	20	39	59	62	79	85	93
17	23	40	60	63	80	86	93
18	26	41	61	64	81	87	94
19	28	42	62	65	81	88	94
20	30	43	64	66	82	89	95
21	32	44	64	67	83	90	95
22	34	45	65	68	83	91	96
23	36	46	66	69	84	92	96
24	38	47	67	70	85	93	97
25	40	48	68	71	85	94	97
26	42	49	69	72	86	95	98
27	43	50	70	73	86	96	98
28	45	51	71	74	87	97	99
29	46	52	72	75	88	98	99
30	48	53	72	76	88	99	00
31	49	54	73	77	89		
32	51	55	74	78	89		

APPENDIX G

Conversion Factors

To convert	Into	Multiply by	Conversely, multiply by
Metres	Feet	3·28	0·305
Centimetres	Inches	0·39	2·54
Square metres	Square feet	10·8	0·093
Sabins (sq. metre)	Sabins (sq. ft)	10·8	0·093
Square centimetres	Square inches	0·155	6·45
Cubic metres	Cubic feet	35·3	0·0283
Cubic centimetres	Cubic inches	0·061	16·4
Kilograms per sq. metre	Pounds per sq. ft	0·205	4·88
Kilograms per cubic metre	Pounds per cu. ft	0·062	16·0
Horsepower	Kilowatts	0·745	1·34

For calculation of reverberation time (formulae 1 and 3, pages 51 and 53) using Imperial units the constant to be used is 0·049 instead of 0·16

Index

Numbers shown in *italic* refer to figure or plate numbers.